FINDING FISH

ANTWONE FISHER
with Mim Eichler Rivas

HarperTorch
An Imprint of HarperCollins*Publishers*

HARPERTORCH
An Imprint of HarperCollins*Publishers*
10 East 53rd Street
New York, New York 10022-5299

The William Morrow hardcover edition contains the following Library of Congress Cataloging in Publication Data:

Fisher, Antwone Quenton.
   Finding fish : a memoir / by Antwone Quenton Fisher and Mim Eichler Rivas.—1st Morrow ed.
     p. cm.
   1. Fisher, Antwone Quenton. 2. Afro-Americans—Ohio—Cleveland—Biography. 3. Foster children—Ohio—Cleveland—Biography. 4. Cleveland (Ohio)—Biography. 5. Cleveland (Ohio)—Social life and customs—20th century. 6. Afro-American screenwriters—Biography. I. Rivas, Mim Eichler. II. Title.
F499.C69 N33   2001
973'.0496073'0092—dc21
[B]   00-031872

First HarperTorch paperback printing: December 2002
First HarperCollins trade paperback printing: December 2001
First William Morrow hardcover printing: February 2001

HarperCollins ®, HarperTorch™, and ❦ ™ are trademarks of HarperCollins Publishers Inc.

Printed in the United States of America

Visit HarperTorch on the World Wide Web at www.harpercollins.com

10 9 8 7 6 5 4 3 2 1

For my wife,
LaNette,
and my daughter,
Indigo.
My entire world.

author's note

This is a work of nonfiction. The events and experiences detailed herein are all true and have been faithfully rendered as I have remembered them, to the best of my ability. Some names, identities, and circumstances have been changed in order to protect the integrity and/or anonymity of the various individuals involved, and especially to protect the orphans and foster children I have known, who have a right to tell their own stories if they so choose.

Though conversations come from my keen recollection of them, they are not written to represent word-for-word documentation; rather, I've retold them in a way that evokes the real feeling and meaning of what was said, in keeping with the true essence of the mood and spirit of the event.

contents

Save the children.

—Marvin Gaye

pre-memoir

AN UNINVITED GUEST

This beginning part is for my father. This is your story. I wasn't there but I put it together from what the family remembered. And from what I dreamed. You may think it's not my place to tell. Maybe so. But aside from your blood that's in me, this story's just about all I have of you. I guess that makes it mine, too.

The year was 1959, the time a Thursday morning, the second Thursday of June. This was in Cleveland—the kind of big midwestern city that made for a good place to raise kids and dreams. The economy was thriving then thanks to the motor companies, the steel mills, tool-and-dye factories, and the other industries springing up across the landscape every which way but north, where the great, ominous expanse of Lake Erie is all that the eye can see.

Cleveland in the 1950s was a proud place, a righteous place. A brand-name city—Ford, Republic Steel, White Motors, Fisher Body, Stroh's beer. A family city. It was work and church. It was a ball town. Football and baseball. Especially baseball. It was music, too. Gospel, doo-wop, jazz, blues, and the symphony. It was the birthplace of Superman.

Cleveland was already a big city, on its way, at one

point, to reaching number five on the list of largest cities in America. But to men like Horace Elkins, Sr., a hard-working father of eight, it still had that small-town, friendly feeling. True, he liked to keep himself and his family close to home, safe in the bounds of the Glenville area. This was the working-class, predominantly black neighborhood near the lake, nestled comfortably between the main arteries of St. Clair and Superior, which were linked by 105th Street, forming an l shape. Up and down 105, as it was called—at the grocers, the five-and-dimes, thrift stores, barbershops, mom-and-pop record shops, clothing, shoe, and liquor stores, and in the neon-lit lounges—residents and merchants were on a first-name basis. On the street, strangers were scarce.

Horace knew and trusted his neighbors; they knew and trusted him. More than trusted, they looked up to him. A man felt good to walk proud among his own. Nothing made Horace more proud than his family. That's why he was forever telling his sons, "Get yourself a rib." Find a woman, settle down, raise children, be a man.

Friendly and small-town it could be, but Cleveland living was hardly easy. Like the weather and the work, the temperament was harsh. And Horace also understood there was an air of danger in the daily hum—something seductive but lethal. He wondered sometimes if that sense came from the Choctaw in him, the way he always felt on guard. When he could, Horace liked to shake those superstitions. They represented the old ways, the primitive beliefs brought up from Arkansas, where he was raised, and the slave ways that went back to the Elkins, West Virginia, plantation where the slavemaster Elkins had passed his surname to Horace's stepfather, who insisted Horace change his name from Barnett to Elkins. They were ways not of living but only surviving that had mi-

4

grated north with his people and stayed with them, even after they left behind the plantations and reservations.

Horace fought the old ways that lurked inside him because he wanted more from life than a subsistence diet of fear and faith. So he fed himself an education, in night classes and on his own, studying the works of Shakespeare and Edgar Allan Poe and the philosophers from throughout the ages. Through philosophy, literature, science, music, and art, Horace Elkins became a learned man, a free man, a doctor of medicine, in fact, and, in spite of the modest apartment on Parkwood Avenue in which he and his family lived, a kind of noble man.

Knowledge, to him, was power and redemption. Knowledge, that is, and the Catholic Church. The Elkinses were strict Catholics. This, too, for Horace, was a further rejection of the old ways. But even so, he couldn't entirely rid himself of ancient instinct. Sometimes, those uneasy feelings had to be heeded.

Rising at 4:30 A.M. that Thursday, as he did every day in order to catch the bus on time that would carry him to the first of his two jobs, Horace woke to the shattering blast of a gunshot. He sat up, waiting for the aftermath. But nothing. A dream, that's all, he thought. Too vague to mean anything.

Horace washed and dressed quickly and went into the kitchen, where Emma had already prepared his breakfast and was setting it out for him on the kitchen table. He took his seat, bowed his head in silent prayer, and then began to eat.

Emma went back to the stove and stood watching her husband, waiting to see if he needed anything. For a moment, she was young again—before nine babies—back in Forest City, Arkansas, looking for the first time at Horace, the light-skinned young man with his Choctaw warrior nose, who bowed as they were introduced, saying, "Hello, Miss Emma." Like him, she was a blend of red and black, Choctaw and African, whose lineage could be traced to the slave plantations of Alabama—and she felt she belonged with him.

She was thin then, a little slip of a girl, but strong, Horace saw; and she was stronger than he could imagine. Over the years, as the lines of her body widened and curved, her position of authority was recognized not only in the family but within the community. Everyone around had heard stories about Emma bringing home other people's hungry children and feeding them right along with her own nine kids.

Emma was strong, and her word was the law. Back in the days when Horace came home every night complaining about his job at the hospital laundry where his white boss cruelly mistreated him, she lifted him up each time, reminding her husband, "But it's a good job, Horace. Look at what we tryin' to build for the family. You stay with it, Horace. Just hang in there. We almost where we need to be."

He'd listen to his wife because when she felt something, she felt it strong, and Horace would leave the next day, girding himself for more abuse. Then one afternoon, hours before his shift was normally over, he came walking up Parkwood, lunch can in hand, whistling. Neighbors out on their stoops and on the sidewalk were surprised to see Mr. Elkins home so early on a workday. They were even more surprised when Mrs. Elkins appeared in the open window of their apartment as he called to her, "You know that good job you love so much at the hospital? You might want to go apply for it. They just got an opening."

The family loved to tell that story—especially the

boys. The two youngest of the four sons, Spinoza (Horace named him after the philosopher) and Raymond, thought it was funny. But Horace, Jr., the oldest boy—already in his mid-twenties yet still living at home—felt differently. To him, there was nothing amusing about a black man having to subject himself to oppression in any form. Now, Eddie, the second oldest, felt that story and what it said about his father gave him justification to walk away from all kinds of situations he couldn't tolerate. Horace, Sr., was none too happy about that.

Emma took Horace's plate back to the stove to fill it once again. He thanked her with a nod of his head, apparently hungry though lost in his own thoughts. Emma went back to the counter to finish packing lunches, thinking to herself of the day when Eddie was eighteen, not long after enlisting in the armed services, that Horace found his son had shown up unannounced on his doorstep.

"Wha'choo doin' here?" Horace, Sr., asked, scrutinizing his slender, good-looking son.

"I'm on leave."

"Well, let me see your leave papers."

Eddie pulled a map from his uniform pocket. It was a map of the United States, folded to a section of the Midwest with Cleveland circled in red. Eddie slapped it down on the table. "There's my leave papers!"

Horace sent his son directly back, but Eddie went AWOL a few more times, even to the point of having a couple of MPs turn up on Parkwood to haul him off yet again.

Before long, Eddie was given a dishonorable discharge and he returned to his parents' house, no worse for the wear.

Sometimes Eddie could be trouble. That early Thursday morning, Emma stood in her kitchen and worried about her son Eddie. The dreamer, the artist, the fighter. Goodness, he had a gift—a singing voice that flowed from him like the angels put it there as a surprise for anyone lucky enough to hear. And a way with words, in speaking and in the poems and letters he wrote. She had a keepsake box filled with his writing. And he was beautiful. All her children were attractive. All tall, all with unusual-colored eyes, ranging from brown to green to silver. But Eddie's eyes, they were something else. Something out of this world.

Emma had heard some parents called him "Swami" because, they said, Eddie hypnotized the girls into doing what he wanted. That worried her. Not because she believed it. But because she understood those parents' fears. After all, she had four daughters of her own to look out for.

Eda, the eldest of her girls, could take care of herself. Intelligent and independent, a nurse no less, she had already married and moved away to Chicago. Ann, her second-oldest girl, kind and warm, was still cautious. That was good. But with the twins, even in their teens, it seemed the boys were much too interested for Emma's liking.

The same was true of Eddie. There was never a time Emma could recall that he didn't have at least one girl coming around for him.

Eddie was charming but he had a mean side, too—an unpredictable temper that could explode on anyone at all. Emma was the only one whose warnings he'd even consider. "Be careful, boy," she was forever telling him, "stay out of trouble."

What she meant was: Eddie, you're different, you're special, I got such plans for you. Of course, all her babies were special, different, and she wanted the best for all of them, education and college; for her daughters safe and loving homes, for her sons better chances to raise themselves up and become men of substance, stature. They all showed promise: Horace, Jr., with his ideas about changing society; Spinoza, ambitious, maybe a little too ambitious; and her youngest, Raymond, always drawing and painting, making up his stories you couldn't half believe but always did because you wanted to.

But Eddie, he was like something extra-magic, like a person who could go out there and win, a person who could really be somebody. Somebody big in the world. An author or a poet.

Emma saw that Horace had finished eating. Should she tell him about Eddie's decision to go to Chicago? She wanted to tell him about Frances, the mother of Eddie's two daughters, and the latest trouble, so Horace would talk to his son. But what to say?

Emma went silently to his side. She filled his coffee cup, took his plate and utensils, wiped down the table. Horace Elkins was a no-nonsense man. If she told him about Eddie's plans to go stay in Chicago at Eda's, where there were more opportunities for nightclub singing, Horace would be all for it. But should she tell him that Eddie was intent on getting Frances and the girls out of the house because Frances's stepfather was being released from jail after serving time for molesting her, his own stepdaughter; not to mention that the man had given her a child? Should she tell him that Eddie was insane over the idea that the stepfather would return and repeat his actions with Frances and Eddie's two daughters?

Emma wanted her husband to put down his foot and make sure Eddie didn't go by there anymore. A month ago Eddie went over with a gun, threatening Frances to take the girls to go live at his sister Ann's. When that didn't work, a couple of weeks after that, Eddie went over again and Frances's stepbrother took out a twelve-gauge shotgun and fired at him, buckshot grazing his pants legs.

What Eddie was trying to accomplish, Emma knew, was the right thing. But he was going about it the wrong way.

No, Emma reconsidered, she didn't want to upset Horace in the morning. Working the two jobs and trying to make house calls on his patients in the neighborhood, he had enough weighing on him. Better to bring it up that evening.

Horace tore himself from his own reverie, glanced up at his wife, and gave her a look: What is it?

"Never mind," she said, and placed her hand on Horace's shoulder as he stood up.

"It's hot already," Horace said at the open door. Emma gave him his lunch can. They said a fast good-bye and she stood there, leaning on the doorpost, watching him leave, aware of the warming air rising from the lake.

Horace was halfway down the sidewalk when he stopped, hit hard again by a wave of uncertainty. He had sat through breakfast, wanting to tell his wife but not wanting to worry her. From the look of her, she already had enough on her mind. No, Horace decided, it was nothing, and he continued along his route to work, away from his home where, after this day, nothing again would ever be the same.

The rest of the Elkins household was now waking up. Soon Horace, Jr., and Spinoza had eaten and gone. After them, Emma laid out food for the twins and Raymond, giving the three enough time to eat and grab their books and lunches before she hurried them off to school. She was cleaning up the dishes when Eddie appeared in the kitchen doorway, grinning from ear to ear.

"Wha'choo grinning about?" Emma asked, unable to resist the contagion of his smile.

"I don't know, Momma." Eddie shrugged. He was dressed in his special black suit. He didn't sit, not wanting to put any creases where they didn't belong. His hair, conked, parted on the side, and slicked back, was immaculate. He stood by the window, catching a breeze that seemed to keep him cool. Outside, all through the neighborhood, clotheslines of laundry hung and swayed in the summer morning air. You could even smell it drying.

Eddie told his mother about his phone call to Eda two nights before. His sister was excited about him coming to Chicago, he said. Everything was looking up.

"That's good, Eddie," Emma said, and paused to dab her forehead with cold water from the tap.

She may have been a country girl, but nobody could fool Emma Elkins. She had what they called mother wit. Horace got his uneasy feelings, but Emma could read the future.

And now suddenly Emma wanted to say something else, something important. She didn't know why. She wanted to tell Eddie everything he needed to know in case she never saw him again. That very thought sent a sudden sick, clammy feeling straight through Emma; it began at her feet and rose to the top of her head, filling her with overwhelming panic.

Emma's eyes peered into her son's. She spoke calmly but only to mask her fright. "Where you goin'?"

"Nowhere."

"Be careful, Eddie. Don't get in no trouble."

"Aw, Momma, wha'choo talkin' 'bout? You know I

don't never cause no trouble," Eddie said, and again flashed that devilish grin. As he did, he went to her, threw his arms around her neck, buried his head there.

She held him tight, as tight as she could, for as long as she could. "You heard me," Emma said close to his ear. She drew back and let her son go.

They walked together to the door. She watched him start down the sidewalk.

"Man, it's hot," he scowled, and then took off.

Emma stayed at the open door, her body growing heavier with every step Eddie took, until he had gone completely from her sight.

She went inside, back to the kitchen, and sat down by the window and looked out at the rows of laundry draped across the neighborhood. For the moment, it seemed to give her some peace.

Bumping along the streets of downtown Cleveland on the crowded, overheated bus, Jess Fisher caught sight of the union hall and reached up to ring the bell for the next stop. A short, sturdy-looking, brown-skinned man, he got up from his seat in the back and made his way down the aisle to the rear exit door.

Today, nothing was going to spoil his good mood. Not the long bus ride, not the heat. Not nobody wanting you to sing their hard-luck blues.

When the bus doors lurched open, Jess hopped down onto the sidewalk and made a beeline for the union hall. Under his breath, he half hummed, half sang the words from "Got a Job," the Miracles song they played on the radio all the time when he was waiting to get out of the service. Now he was out, back in Cleveland, and, yes, indeed, after not too much hunting, had got himself a job.

Not just any job. This was a union construction job. So here he was, walking into the main union office to pay his first dues, with an hour to spare before he had to show up at the work site.

The union office, with its high ceilings and ornate moldings, was just like he imagined it—big, important, loud, busy. At a front information desk, a white man in rolled-up sleeves with sweat stains under his arms sat in front of a pile of papers, sifting through them, smoking a cigarette. When Jess mentioned the name of the person he was supposed to see, the man waved him in that direction. He was in. Just like that. Walking past the pearly gates. That's how he felt, almost, like if he died and went to Heaven, God would be a union man.

Jess went to the appointed office, filled out his paperwork, paid his money, and then sailed back out to the street, a member of the club. He arched his head back, looking up at the Cleveland skyline, admiring the hometown he'd once escaped, glad to be back. Like those modern monoliths being built, his future rose before him: steady work, marriage, family, a home of his own.

Leave it to the Fish Man, Jess congratulated himself, remembering how his buddies in the service used to think he was so resourceful. Maybe he was. But only because he had to be. Five years before, after his momma died, he was left, the oldest of six children, to deal with a daddy who drank and beat all of the kids, but him the worst. There was a war in Korea, Jess was old enough to join the service, and so he enlisted, figuring it was either that or kill the old man.

While he was away, he had a letter here and there from various relatives but for the most part lost touch with his brothers and sisters. It seemed as if they had dispersed like refugees from a war zone, some faring better than others. Jess hadn't tried to contact anyone yet. He wasn't ready. First he had to get himself settled and stable. Then he'd see about the others. Mainly it was Eva, the youngest, he was concerned about. She was only twelve or so when their momma died and had been left alone with their father without any protection. Then the court deemed the old man unfit to be a father, and Eva was forced into the foster care system. That was another hell, Jess knew.

He heard about some trouble. At first Eva stayed at a foster home on the east side in the Glenville neighborhood. She went to school, hung with a bad crowd, ran away from the home, was taken to juvenile court for truancy and something else, he wasn't sure. There were other "episodes," as the social worker called them. He heard she got sent to Girls Industrial School in Delaware, Ohio, about three hours southwest of Cleveland.

Jess thought of little Eva, with her dark complexion and the long, thick hair she was so proud of, lost in her own world most of the time. She had a real nice singing voice, when she let you hear it. Mostly, she sang quiet to herself, as if music carried her to a better place, somewhere else that she could dream about. From what he remembered, his sister wasn't resourceful, not in the least. But three years in prison had a way of changing that. Not that she'd been in the whole time. According to her friend Evelyn, she'd been to Cleveland on a few visits.

When Jess ran into Evelyn some weeks before, she told him that she'd been meaning to get down to see Eva. Not the shy type, Evelyn raised her eyebrows to suggest he was welcome to join her.

All Jess offered was, "Well, tell Mae-Mae I'll be by to see her as soon as I can."

Thinking back to that conversation as he climbed onto

his bus, he really hoped, for Eva's sake, that Evelyn had made it down to Delaware to visit his sister.

Earlier that morning, Eva Mae Fisher had been hoping the same thing. At 11:23 A.M.—that's what the clock in the visiting area said—she knew Evelyn wasn't coming. Again. Oh, well, it was too awful hot to cry. Eva stood up from the visiting table, with effort, and began to lumber away. Around her, other girls sat with parents, boyfriends, husbands, friends, everyone in close huddles. No one appeared to notice her there by herself. Didn't bother her. Noticed was special, she'd learned; special was dangerous.

Seventeen years old, seven months pregnant. It was a fact about Eva that seemed not to surprise anyone. She could still see the ho-hum looks on the social workers' faces at Child Welfare when a routine checkup revealed she was four months along. "This is what happens . . ." the one social worker said as she drove Eva back to Girls Industrial. The statement hung in the air like that, half said.

Eva wondered just what it meant. She wondered why the social worker, a bird-looking white woman in her late twenties, had got so tired already. Eva glanced at the woman, thinking, How she know? Like she expected it, like she saw it all the time.

They all talked like that, no matter what your problems—getting beat up by your daddy or at the foster home; not being able to find another foster place; getting caught shoplifting; skipping school after some boy called you crazy and some girl made fun of you, saying, "When you gone get rid of them teenager bumps?" That's the problem, the social workers say, you're a teenager, a teenager committing legal offenses. After two years in lockup, sixteen years old, you come back to Cleveland and get another chance to prove yourself to a local foster home. But no, you got to go and get pregnant and then, they don't have no choice, it's back to lockup. This is what happens.

And then the social worker had to ask, "Do you know how this happened?"

Eva had been staring out the window, watching scenery change as they drove into more rural Ohio areas. It took a second for the question to register. "How *it* happened?" Eva didn't hide her resentment. "Yeah, I know."

"Does the father know? Do you know who the father is?"

"Unh-unh." She shook her head no. It was a half lie. No, the father didn't know, but yeah, Eva knew who the father was. It was her secret. Nobody could make her tell, not if she didn't want to. Except maybe, Eva decided then, she might tell Evelyn. As long as she didn't ask all the other questions: Do you plan to give your newborn up for adoption? If not, how will you care for your baby? What work can you do? Who will help you?

Walking down the cinder-block corridor of Girls Industrial School on that Thursday, heavy with child, heavy with heat, Eva didn't yet know what her plans were. She'd been hoping Evelyn could come up with a plan. That was Evelyn. She always knew everybody, knew where to find anything. Evelyn was the one who introduced Eva to Eddie in the first place.

It was on 105, a few days before Thanksgiving. Starting to get real cold out already. Eva had seen Eddie a few times before. She'd heard about him before that, too. He was the big brother of the Elkins twins, she knew. He had a group called the Jive Kings. The name reminded her of some other high-class music names—Duke, Count, the

Royales. Everybody said he sang just like Tony Williams of the Platters. She knew they called him "Swami," but she didn't know why. Then she saw him for the first time. He was coming out of the barbershop with a do-rag on his head, pulling a brim hat on top of it. And he looked out from under the brim right at her. For a split second he stood there about to say something, and Eva felt the street spin as he turned to go the other way up to the lounge on the corner, walking like a cat moves with a fast long-legged spring. She liked him from that moment on.

Another time, Evelyn and Eva saw him standing outside the record shop singing with another guy. Evelyn pointed to him and said his name was Bobby Womack and he was getting a record deal. Eva listened. They weren't singing anything in particular, just playing around, harmonizing together, but she thought Eddie's soulful falsetto was even better than Tony Williams's.

That's exactly what she told him the first time they spoke. It took two days of begging Evelyn to introduce her. "Don't getcha hopes up, Mae-Mae," Evelyn warned. "He got a lotta girls."

But Eva persisted until Evelyn gave in.

Eddie grinned when Eva complimented his singing. "You like the Platters, huh?" he asked, and she named some of their songs. They stood on the sidewalk in the cold November air, talking about music and singers, and Eddie looked impressed. When he asked how old she was, Eva lied and said, "Eighteen."

It was a fairy tale. Like nothing that ever happened in her whole life. How they talked together and walked through Dupont Park over amber leaves that crunched beneath their feet, swung on swings, laughing till tears that seemed to wash away their problems. They stayed out all night and went to blue-light-in-the-basement and rent parties, drank cheap wine, smoked cigarettes, and danced the slow drag; he was her prince and she was Sisterella.

Eva heard he was slick. She knew he had other women, one, Frances, they said he was going to marry. He had a child by her, they said, and another on the way. But he didn't stay with her and it gave Eva hope. She heard he was mean and that once he even took his own brother Spinoza's new chrome gray continental suit and sold it, claiming he needed the money. She heard he ran into his youngest brother, Raymond, doo-wopping on Parkwood and Primrose with some other knuckleheads, doing his best to sound like Tony Williams, and Eddie, furious that Raymond was infringing on his thing, slammed his grip around Raymond's throat and began to choke him, his little brother.

Eddie could be unpredictable, but he came from good people. She saw that when she went by the Elkins home looking for him one day. Important people, they must be. Books and paintings all around. Now that was something.

To Eva, Eddie was gentle and patient. Walking outside with the year's first snow on the ground, he noticed Eva bundling herself to warm up, and he led her into an apartment building near the park, where his friend had a place and often let him bring girls. A small one-room with yellow faded walls and a rickety army cot, it was just as cold in there as it was outside, and she didn't care. Only that he laid down his coat for her and took her into his arms.

That was seven months ago. Winter was over; it rained hard all through spring and now summer burned.

"We ain't together," Eddie had said, using the excuse that Eva was too young.

Evelyn reminded her, "I told you how he was."

This is what happens, Eva thought; fairy tales are just that—fairy tales. She heard the noon bell ring for lunch and went, slowly, toward the cafeteria to take her place to wait in the long line of girls murmuring to one another. She knew some of them, knew some of their stories. None of them knew hers. Some maybe felt sorry for her. But she didn't feel sorry for herself. She had a plan. It came to her right then. She would have this baby and then go see Eddie and he'd find her a place to live and make sure she and the baby had everything they needed. He was going to be rich soon anyway. In fact, she was going to write him a letter in a few days, so he could start getting things ready. Thinking about Eddie and how one day they could even be together for real—she wouldn't be too young anymore—made her smile. She stood in that line grinning with her secret and her plan, feeling the child inside her flipping and turning and kicking, happy as can be, like he was dancing to the tune of "I'm in Love Again," her favorite Fats Domino song.

At five minutes after twelve noon, Eddie stood downstairs at the house on East 90th Street where Frances's mother had a place upstairs. His black suit jacket off and folded carefully over the railing of the exterior wooden staircase, he paused on a lower step to mop his forehead with a handkerchief, loosen his tie, and unbutton his shirt collar. For a second time, Eddie hollered up to the open bedroom window where he knew Frances and his two daughters were, even though she refused to answer him the first time he called her name. But no answer again.

Eddie wailed with his third try and began to climb the stairway. On the landing, he waited. Again no answer.

Inside the bedroom, Frances paced the small area, staying clear of the window and the chain-locked back bedroom door that Eddie was now approaching. In one arm she held her infant daughter; her other hand tugged at the elbow of her two-year-old, who was crying and trying to pull away, saying, "It's Daddy! I wanna see him. . . ."

"Be quiet," Frances whispered, taking the toddler and the baby to her mother's room, pointing her finger at the oldest child to stay put. She took a deep breath as she heard Eddie banging and kicking on the door, yelling, demanding that she come open it.

A pretty, soft-skinned nineteen-year-old, she wore a torn housedress already drenched by the heat and fear. Frances could feel her insides at war: stomach twisting, heart slamming up into her throat; her brain seemed to be swelling tight against her skull. Everything inside begged for release from the voices yelling on top of the other, telling her what to do, drowning out her desire—to just go with Eddie no matter where and do whatever it was he wanted her to do, because she loved him so.

But no, she had tried that, the voices inside mocked her; Eddie was hell on wheels and if he had put his hands on her before, he might do it again.

Frances could hear her brother saying to her, "How come you say you love 'im after he beat you up? It don't make no sense." She had argued with her brother, telling him, "He jes' mad . . . you don't understand him."

"Open this fuckin' door!" Eddie yelled, giving it another hard kick.

Frances had to be firm, she decided, and let Eddie know he couldn't push her around anymore. She went to the door, put her back against it and turned her head to shout, "Go on, Eddie, you know you don't supposed to come 'round here."

"Fuck that. I don't supposed to come around here? I heard your daddy's getting released from the joint next week!" Eddie's voice softened as he continued to talk through the closed door. "Now, you know me. . . . You better come on now, girl. You-all got to come with me. My sister said you and the girls can come back and stay with her." He paused. When she didn't answer, he exploded again, "I ain't playin' around no more, Frances! Get the kids and your shit and come on."

"I got a peace order on you," Frances finally replied.

"A peace order? I don't give a damn about no peace order. Like I need to be restrained? Open this fuckin' door." With that, Eddie kicked the door one more time, breaking it open as far as the distance allowed by the chain lock, which prevented it from swinging open all the way.

Frances hurried to the closet in her mother's room, where she got her mother's shotgun, telling the children, "Don't worry, Mommy'll be back."

As she exited the room and started back to the door, she could hear Eddie still ranting about the peace bond.

"How you gonna put a restraining order on me? What about that child-molesting daddy of yours! You take one out on him? I ain't gonna have that nigga peeking in my babies' diapers! Shit. Now, I'm the villain?"

At the door, Frances raised the gun, sticking the barrel out past her chain lock, through the narrow opening.

The sight of her in that pose, threatening him, the man who loved her and was there to protect her, made no sense. It was almost laughable. That wasn't his Frances. There wasn't a mean bone in her body. Eddie went to kick the door again, intent on breaking the chain because it was part of the script he knew about who he was— because nobody messed with him and if he had to bust down a door, he would. And just then she pointed the gun, aiming it right at his stomach, and in that hairbreadth of a second, just before his shoe made contact with the door, hard enough to break the chain lock and allow the door to swing wide open, Eddie had the shocking realization that the script was different this time. It was too late to stop himself from kicking the door wide open, too late to stop himself from gripping the barrel of the shotgun, too late to keep Frances from pulling the trigger and releasing an explosion of fire straight into the center of his body.

Screaming with the horror of disbelief, Frances watched as the impact of the buckshot struck Eddie's stomach, practically severing him in two. His body flew up, back, and then downward without hitting a single stair step, crashing to the ground with an earthen shudder. As Frances ran down the steps after him, the two-year-old came to the open door and looked down at her mother, in tears and kneeling by her father's side. She heard her daddy look up with a frozen expression of surprise and utter his last words to her mother, "You killed me."

As the story goes, a woman of color who worked at Mt. Sinai Hospital on 105 was just getting off the early shift when they brought in Edward Elkins to the emergency ward. In 1959, in Cleveland, there was nothing commonplace about the murder of a young man in the 'hood—what they called the ghetto or the slums back then, back before it sounded nicer to call it the inner city. This woman, she didn't know who the twenty-three-year-old was. But the sight of him was so devastating that when she stopped by to visit a friend who lived in the 900 block of Parkwood, she couldn't shake it from her mind.

As she was telling a group of neighbors about the handsome young black man and how they carried him in nearly cut in half, but still alive, she described how he fought not to leave consciousness. He had the strength and will to live of ten men, she swore. In that group of neighbors, so the story goes, was Emma Elkins, listening to the tragic details about his tortured death, unaware that it was her own son. Emma must have made the sign of the cross and said a silent prayer for the mother, who-ever she was, for God to take pity on her and spare her whatever pain He could. Only a short while later, with the arrival at the Elkins apartment of someone from the neighborhood who had been at the hospital and had recognized Eddie, would Emma know she had been praying for herself.

The funeral was the following Tuesday. It was said that Horace Elkins, Sr., broke down and sobbed like a baby—a man who had never in his life been seen to cry by any one of his children, not even many years before when he lost his young daughter Teresa. It was said that Eddie's death virtually killed Emma and that her health began to decline. She never spoke about what had happened. In fact, she instructed the entire family never to discuss the way her son had died and never to speak ill of Frances. Those two girls were Eddie's daughters, Emma explained; they, and their mother, would be forever treated as a part of the Elkins family and not made to suffer any more than they already had.

Emma's word was the law. Everyone obeyed. If they hadn't, they would have had to contend with Spinoza, who eventually became the family law enforcer. When Emma died, it was Spinoza who took the trunk of family keepsakes with Eddie's poems and letters and locked it away, never to be opened again. The rest of the family had memorized some of the poems; they would recite them on special occasions and, in time, talk about the happy, funnier memories of Eddie, but never about the trouble to which his own actions had contributed, trouble that ended his life, so full of promise, at age twenty-three. All of that was virtually buried for good the day they lay his body in the ground of Calvary Cemetery, in a grave unmarked except for a small placard that bore the burial plot number.

On June 14, 1959, the *Cleveland Call and Post* ran a story with Edward's picture and reported the local prosecutor's ruling that the homicide was justified. It was the last official mention anywhere of the life of Edward Elkins.

Two months later came the first recorded mention of me and my life:

```
Ward No. 13644. ACCEPTANCE: Accep-
tance for the temporary care of Baby
Boy Fisher was signed for by Mr. Nesi
of the Ohio Revised Code, Section
5153.16 (G). CAUSE: Referred by Di-
vision of Child Welfare on 8-3-59.
Child is illegitimate; paternity not
established. The mother, a minor, is
unable to plan for child.
```

The report went on to detail the otherwise uneventful matter of my birth in a prison hospital facility and my first weeks of life in a Cleveland orphanage before my placement in the foster home of a Mrs. Nellie Strange. According to the careful notes made by the second of what would be a total of thirteen caseworkers to document my childhood, the board rate for my feeding and care cost the state $2.20 per day. When I was two months of age, the caseworker wrote that I was growing and developing nicely and, all in all,

> Baby Boy Fisher is doing beautifully
> in the foster home, where he is re-
> ceiving much love and security. The
> foster mother commented, saying she
> thinks that he is spoiled.

Over the course of my first year of life, four other visits were documented, with these observations:

> Antwone crawls all over the place,
> stands and walks holding onto
> things, he says hi and bye bye. . . .
> He likes pancakes and mashed pota-
> toes. . . . He smiles a good deal and
> appears to be a well cared for child.

Though I like to think that my first foster mother's loving influence during my first two years of life had a positive impact on me, it seems that even before my first birthday she began to feel terrible reluctance about her growing attachment to me. At that time she called the child welfare home-finding department to say that she wanted me removed and that she wanted a child around four years old, because she didn't want any more infants. Then she contradicted herself, according to the report:

> She said she wanted him and how can
> he move to a new home, repeatedly
> saying she did not want to give him
> up. . . . I do hope we can straighten

out his situation and re-evaluate his
natural mother. At present she has
shown almost no interest except by
an occasional phone call. . . .

Not long after my second birthday, a visit was recorded
that clearly showed the level of my attachment to my foster mother, noting that I refused any attempts to be lured
away by toys and distanced from her. The caseworker
even remarked that when she entered the lobby,

The foster mother was sitting with
the child and she had him held in her
arms as one would hold a 3 month old
infant. Child is large for his age
and is a very attractive baby. He has
even brown coloring, fairly regular
features, and is sturdily built,
lying in her arms contentedly like
an infant. Before the foster mother
left the office, she was admonished
about holding the child so close to
her, making such a strong emotional
tie between them . . . making it that
much harder because the separation
would have to be made regardless of
her feelings for the child . . .

In September 1961, the caseworker began to seriously
explore foster homes for me. This resulted in a decision
that I be placed in the home of the Reverend and Mrs.
Pickett, a couple in middle age already with grown children of their own, children from previous marriages, and
grandchildren:

Mrs. Pickett has had a large family of her own and is not easily threatened by rejection in children. Mrs. Pickett is a product of the rural south which is evidence of her close ways, expressions and her relaxed easy manner. Caseworker attempted to explain Antwone's close tie to his present foster mother, his fear of strange people and the need at this time to share him with his mother. Mrs. Pickett has a feeling of confidence in winning children and did seem to have a bounteous amount of patience and fortitude.

On October 4, 1961, Mrs. Strange took me for the first of three visits that week to the orphanage, with a plan under way for the transfer of my temporary custody:

Caseworker had colored paper and crayons for Antwone to experiment with. His face lit up at the sight of this material and he enjoyed these things. . . . CW showed him how to use the different colored crayons and on different colored sheets of paper and he relaxed a great deal.

Two days later, I was brought in again. But this time, the report says, when they showed me the art supplies I only wanted to take them home with me and refused to leave the chair where Mrs. Strange was sitting. Three days after that, the caseworker noted that I was not in the least interested in the material and clung tightly to my foster mother. When they advised her that on the next visit she would have to leave me there, it was said Mrs. Strange became very upset, saying she couldn't go through with it, begging them to come to her house to take me. Instead they came up with a different plan:

On 10-13, foster mother's friend brought Antwone in from their car. Also her little adopted son came into the agency lobby with Antwone. . . . They arrived at the door to the lobby and the friend and the older child quickly slipped back out the door. When Antwone realized that he was alone with caseworker, he let out a lust yell and attempted to follow them. Caseworker picked him up and brought him in. Child cried until completely exhausted and finally leaned back against CW, because he was completely unable to cry anymore.

When I was later taken from the orphanage to the Picketts' house, I cried for another long while and refused to remove my coat or allow anyone else to remove it. At that time, I wouldn't let go of the caseworker, and it was only when I was taken down into the playroom in the basement by "another small ward" and became interested in a bike that the caseworker "slipped away." All through my case files, everybody always seemed to be slipping away, in one sense or another.

Of course, none of this stayed in my conscious memory. Nor did I retain any reminders of the circumstances of my birth or information about the many hands that had passed me along, from one to the other, to where I found myself growing up as a part of the Pickett household. For all intents and purposes, my life begins then and there— just before the age of four, a time when I can recall my earliest memories.

At first, I wasn't told anything about being an orphan or a foster child. Even though everyone in the household had a different last name, which was confusing, to the best of my understanding, the Picketts were my parents and the other children of varying ages were my brothers and sisters. But for all that I didn't know and wasn't told about who I was, a feeling of being unwanted and not belonging had been planted in me from a time that came before my memory. And it wasn't long before I came to the absolute conclusion that I was an uninvited guest.

It was my hardest, earliest truth that to be legitimate, you had to be invited to be on this earth by two people— a man and a woman—who loved each other. Them just having sex wasn't enough to get you invited. Each had to agree to invite you. A mother and a father.

At the time that I realized the Picketts weren't my parents I came up with the idea that some awful hospital mix-up had taken place and that my real mother and father were looking for me and would find me at any minute. Then I decided they actually knew where I was and for reasons I hadn't figured out exactly, I couldn't be with them just yet but that was only temporary. And in the meantime, they were keeping close tabs on me. My frequent visits to the clinic at the orphanage, for example, I saw as just a setup for my mother and father to catch a glimpse of me. I knew one day, any day now, they'd introduce themselves to me and that would be it. They'd take me home with them that very fine day. So I paid attention wherever I went, in case I spotted them before they did me. I wanted to be on my best behavior so my parents would be pleased to claim me.

At the clinic, the supermarket, the barbershop, wherever, I kept an eye out. I played that game with myself that kids play if they see a car that's real nice. One of us would say, "That's my car." And then another kid might say, "That's *my* car." And the first kid would say, "Well, I got that car before."

In my version, I go to my clinic appointment and I pick the prettiest black woman who walks by. That's my mother, I say to myself. Two minutes later, a prettier black woman walks by. That's her, she's got to be the one, not that other woman.

I played this game everywhere and kept switching up. My mother got more beautiful by the hour. Not only beautiful but kind and warm and smart. Now, my father, he had to be the coolest. Cool as in together. Like the men I saw at the barbershop and the drugstore—Mizz Pickett called them "cooties." These were the black guys with the processed hair—conkaline, as it was known—and who, throughout the week, wore a scarf on their head, underneath a brim hat. The coolest guys would wear their stovepipe pants—real slick, sharkskin—with a V-neck sweater that had the V part covered in and, to top it all off, a leather coat. Then some of them even had the James Brown type of short black boots with elastic at the ankle and pointy toes. That was my father—one of the coolest guys with the hats on.

As I got older, I thought less and less specifically about my mother, and more and more specifically about my father. In my mind, he grew to legendary status. He might be a famous sports figure one day or a singing star the next.

30

So I began to search for the words I would say to my father when finally we met and spoke. I saw us walking together, laughing. I'd say, Man, it was rough out there. But now I'm home, I made it. And he'd put his hand on my shoulder and I'd know, without him saying a word, that I made him proud. That I was the son he always hoped for, the one he would have invited, if he could have.

We'll give Antwone no sister to smile
And say to him, when I grow
You must know, your name is my first child's.

We'll give Antwone no brother to share his
   childhood days.

We'll give Antwone no mother, he'll go
   through life this way.

We'll give him no clear vision to see
   his way through strife.

We'll give Antwone no father to help
   guide him through life.

No, what we'll give to Antwone,
From time to time in life,
To find his own, all on his own . . .
We'll make this Antwone's life.

act one

WARD
OF THE STATE

one

If I follow the path of memory back to its start, I begin life looking out my upstairs bedroom window. It's here I have my best daydreams and where I can make up stories I like to think about. In my mind's first flash of light, I am here, on the inside looking out of the Picketts' two-story house on a street at the edge of Glenville, the second house from the corner, a block from 105. This is a snow-covered morning when the other kids, already school age, are gone and I'm alone, staring out into the blinding whiteness, thinking it's no fun being left behind, no one to play with.

There is something about being at this window that makes me feel safe. Depends on the smell, though. Young as I am, I have already learned to tell what kind of day it's going to be by the scent of the air in the morning. I can smell rain coming. Not just rain and weather and snow, like now, but other clues. Pancakes on the stove, I know it's going to be a good day. The smell of eggs and grits or water steaming off the driveway after Mizz Pickett hoses it down, either one means I better be on the lookout all day.

I squint my eyes real hard and try to use my special heat-ray vision to melt all the snow. Nothing happens.

My powers must need more practice, I decide. Maybe I should try the looking-through-walls trick. That way, I can catch everybody in their alien monster faces.

Yep, everybody in this house, except me—aliens. Mizz Pickett, the alien leader, Reverend Pickett (her trusty sidekick), and their older kids, whose last names are Pickett. Even Dwight and Flo, who have different last names, like me. Children aliens. They just have to pretend to be scared, so I think they're human like me.

The thing about aliens is that when I'm not in the room they don't have on their human faces. And they have kind of goat bodies with hooves and horns and Devil bugging-out eyeballs and long black sideburns. But just before I come into a room, they slip into disguise so I can't catch 'em. One day, I tell myself, I'm gonna be reeeeaalll quiet and tiptoe down the stairs from the bedroom and sneak soooo carefully into the kitchen and catch Mizz Pickett standing over the stove in her alien body and face—before she uses her powers to see me first. Cooking raccoons always weakens her powers.

But instead of trying to catch her this day, I keep standing at the window, practicing my snow-melting skills. Right now, I know, if I really go out there, the snow will freeze me, so I better just stay in here and daydream some more.

It seems the time came for the visits to child welfare whenever I was doing something I didn't want to be interrupted from, like daydreaming. That's the first thing. Next, I got to be on the watch that she's gonna try to be nice to me. But it don't never last, and that's why I rather she stay her regular way—mean.

I'm at my spot, looking out the window, and I feel her

there, standing behind me, all in her monster face and stuff, just waiting for me to turn around so she can practice putting on her human face. That's how monsters play. But I don't fall for that 'cause I know she's not nice. Besides, she's too old to play with kids. So I keep staring out the window, pretending I don't know she's there at the door.

Dwight is in the closet. He's mad at everybody 'cause he's gotta go with me on the visit to the social services office. That's where all the white people are. Except the one they call my caseworker, who's colored like me. There's another one I saw there, too. She was late I heard them say. Another time they said she was coming but she didn't. But I don't care. I only like to see the toys they got in the toy box and play with them for a little while.

Usually I have to go there by myself but today Dwight's coming, too, and I'm glad. It means I'm not the only one wearing the church clothes. It makes me feel special when I'm the only one wearing them on a weekday. I hate feeling special.

I can feel her behind me, opening her mouth, showing her big sharp teeth, and now I'm scared, but this time, I turn around real fast and she's changed human again. Standing there in the doorway, smiling that fake smile.

"Where is Dwight?" she says.

The closet door creaks slowly open and out steps Dwight, his red-green-and-yellow-plaid Sunday-school bow tie slightly crooked.

"What-choo doin' in dat dhere closet? Nigga! Get on out here so we can get ready ta go! Ya think we got all day?"

"No, ma'am."

She grabs him by the shoulder and jerks at the bow tie to straighten it.

Then she turns back to me with that same put-on smile as I hop to the floor in panic. Panic that she will yell at me next. But instead of yelling, she talks in a making-fun, teasing way, telling Dwight, "Twonny has ta see his momma today." Barely concealing her disapproval of the visit, her mouth twists as she talks and she pushes up her glasses from slipping down her nose, like she's mad at the glasses that she has to take me downtown in the rain.

Her words stick in my ears and my mind begins searching as usual. Mom? Momma? I don't like that, either. It's confusing, hearing about that lady. Who is that? What does it mean? It must not be good, 'cause if it was, I wouldn't hate hearing I gotta go visit her. Momma. The one Mizz Pickett looks at me and talks about, saying, "And *you* and your no-account mammy." Makes me feel bad, 'cause whoever she is, she's too poor to get a bank account; and ashamed that somehow it has something to do with me.

As Mizz Pickett takes out our good coats and commands us to put them on, all I can think about is that it's not fair to make it my fault that some lady I don't know is my no-account mammy. So I don't like her and I'm not gonna like her and it's her fault. And I sure am having a hard time trying to look excited about going to see her now.

In our coats, Dwight and I hurry down what we call the lynoleum stairs, slippery plastic with nailed metal strips on the edges. Somebody somewhere must have had the idea this was safe for children. Besides the Pickett twins, who are teenagers, there are three of us younger kids in the house—me; Dwight, a year older than me; Flo, who's eight years old—and we've all fallen down these steps.

At the lower landing, Mizz Pickett waits in her gigantic fox-fur coat. I know it's fox fur 'cause the head of the unlucky animal is still attached to the collar. Its beady eyes always seem to stare right at me. Mizz Pickett stands there frowning in this furry coat and on her head is her cashmere pillbox hat with a long peacock feather sticking out into the air—*platow*.

These are her *goood* clothes, what she wears to church and when she wants people to think she's an upper-class Negro, unlike, she says, "most all the low-down niggas in the city." And then she always adds, "Niggas ain't nothin'."

At the bottom of the steps, she gathers us in front of her, examining us as if seeing us for the first time ever in her old life.

Then it begins. I brace myself, close my eyes, shrug my shoulders. It stinks. Now I feel it. It's wet and slimy like always. It has to be after she stuck her finger in her mouth, sucked it till dripping with enough spit to wipe the corners of my eyes, nose, and mouth.

She turns to Dwight, who hasn't said a word yet, but now, wiggling to break free from her, begs, "I already washed my face!"

"Shut up, nigga, and hold still 'fore I knock your head off," she replies, jerking him closer. She slimes him, too.

I crack a grin at Dwight. He gives me a mean look in return but then cracks a smile, too. We both knew since last night it was coming.

Mizz Pickett grabs her black patent-leather purse and transfers a few items from her everyday pocketbook into it and she's ready to go. There's nothing dainty about this full-figured, brown-skinned, middle-aged woman, but she holds the patent-leather purse in a real dainty way on her wrist with her hand turned up. I think to myself that this must also be meant to add to her look of a woman of high social standing.

Walking single file behind her, Dwight and I follow Mizz Pickett through the living room, onto the enclosed porch, and out onto the front stoop. The scent of Mizz Pickett's spit still is in my nose, making my stomach sick as the cold heavy raindrops splatter on my face and the wind whips around me. Dizzy and afraid of what's coming next, I feel the need to throw up, but if I do, it'll mess up my church clothes and then I'll really get it. Like before. So I think about something else until the feeling goes away.

Soon, we're all in Mizz Pickett's white Buick. Me and Dwight in the backseat, bumping around and sliding as the car skids along the wet streets. Mizz Pickett is perched behind the steering wheel, both hands gripping it, her head jutting out over it with her chin an inch above and her face almost pressed against the windshield as her glasses slip down her nose. It's funny to me, even now; this is the way she always drives. And it's odd how the sound of the windshield wipers is the beat that keeps her tune as she sings at the top of her voice: *"Jeezus, on the main line, tell 'im what-choo wont . . . jus' call 'im up and tell 'im what-choo wont . . . ,"* singing all the way downtown.

At the front door of the social services office, I notice Mizz Pickett looks like she's getting light-headed, something that happens when she thinks she's gonna get the approval of the people inside. She pauses. Then she adjusts her purse, pulls the door open, and shoves Dwight and me inside, in front of her.

In the lobby, she starts talking all loud, like she knows everybody. "Ha'ya'll doin' ta-day! Ain't God a good God, 'cause he sho is blessin' me!" Then she throws her hand in the air, waving it from side to side, and suddenly points to the heavens as she says, "Touch me, Lawd!" She pauses, waiting for a reply, and then demands, "Right now, Lawd!" *Bam!* As if struck in the kidney, she arches her back, throws her head to the side, and hollers, *"Ohh, glory!"* and begins speaking in what she says is "other tongues." Mizz Pickett has explained to us before that she doesn't know what she's saying when she's using the other tongues, 'cause, she says, "De Lawd ain't seen fit ta bless me wit' de power of interpretation yet."

While she waves and talks in tongues, Dwight and I hurry over to the seats and sit with our hands in our laps and our heads down. Out of the corner of my eye, I can still see everybody in the room looking at us like we're funny, not a laughing funny, but a weird-people funny.

"Mrs. Pickett," says the young white lady behind the desk, in a voice that shows she just remembered the only way to get Mizz Pickett to sit down and shut up is to just tell her.

Not knowing she's about to get molded (what I call being put in your place), Mizz Pickett immediately stops preachifying and comes to attention.

"If you'll please have a seat, Mrs. Pickett, the case-worker will join you shortly."

Whoosh. Mizz Pickett's face goes blank of expression. She lowers her head and replies, "Yes, ma'am." Fidgeting with her pocketbook, she sits down next to us and pushes us both back in our seats, saying, "Sit up, what's ill ya? Ya'll young'uns, ya just so slouchy." Now we're molded.

As Mizz Pickett swings her head in the other direction, her hat's peacock feather just barely misses hitting the woman seated to her other side.

My worker, Miss Jenkins, comes into the waiting area. She's smaller and younger than Mizz Pickett and dresses like the white people. And she's good friends with Mizz Pickett so she greets her first, then me and Dwight. But it is my hand she takes and directs her conversation to me, saying, "Antwone, you've gotten so tall since the last time I saw you. How old are you now? Five? Six?"

Everyone is staring at me. Poor Miss Jenkins must have a bad memory, 'cause it's the same thing she asks every time she sees me. Trying not to make her feel bad in front of everyone, I hold up four fingers to her.

"Sho' he's four!" Mizz Pickett announces to the whole lobby. "And growin' up niiice." Before the floor can swallow me up in embarrassment, Miss Jenkins gestures for us to follow her into the visiting room. I rush to get there first, thinking about how excited Dwight's gonna be when he sees the toy box and trying not to listen to Miss Jenkins as she comments to Mizz Pickett, "He's still not very talkative, I see."

"Well, he sho' runs his mouth at home. We can hardly keep him quiet."

Me? I think. What's Mizz Pickett saying that stuff for? She always asks me, "Why you don't talk? Ain't ya got nothin' to say?"

The question flies from my mind as soon as we enter the visiting room and I see the toy box in the corner—just the way it was the last time, stuffed full of balls, trucks, building toys, books, everything. I look at Dwight and say, "I told you." We run and pull out every single toy until he finds a fire engine he wants to play with and I decide to build a tower with all the blocks.

Miss Jenkins, folder and pen in hand, and Mizz Pickett, purse hung from wrist, stand in the doorway talking.

"How is he doing?" asks the caseworker in a serious, secret voice she must think I can't hear.

"He's doin' niiice. Comin' along real good." Mizz

Pickett pauses and then adds, "He's doin' well with the cleanin'." Nothing makes her happier than having a child of four and a half do a day's worth of cleaning; in my sub-conscious the idea is planted that if Mizz Pickett could have said we children were picking our daily quota of cotton, it would have sent her well over the moon and splashing into the Big Dipper.

After the caseworker makes a note in her folder, she tells Mizz Pickett, "If his mother doesn't arrive in five minutes, we will want to discontinue these office visits. She missed the last two appointments, and her failure to cooperate isn't fair to the child or to you."

"Well, to tell ya de truth," Mizz Pickett says in a low-ered voice, "I ain't never said nuthin' befo', but I never thought she was real bright anyway." Now she whispers even softer, "Anyhow, Antwone, he don't know that's his real momma."

Different, confused feelings start to bubble up inside me, but I turn my attention away from their words, away from my tower, to my favorite robot toy. When I look up, Mizz Pickett and Miss Jenkins are gone and the lady they call my mother is sitting on the couch looking at me.

She's not even a lady, but more like a girl. She's dark like me and pretty, with pretty hair. I look away back to my robot, but I can feel her eyes on me. It feels heavy and it's getting to be too much to bear. I wish she'd just look at Dwight for a while. But instead of getting my wish, Dwight decides to bother me—like he does sometimes—and he reaches over and takes the robot from me.

The mother jumps up and snatches it away from Dwight, saying, "Give it back to him!"

I wait for Dwight to say something smart back, like he normally does, but he doesn't say anything. I know it's 'cause she scared him. I'm glad. She comes over and gives me the robot with a nice smile for me and over her shoulder throws a mean look at Dwight. Then she goes back and sits on the couch. I look at her again and study her face, thinking to myself, she might be nice.

This visit with my mother was different from the preceding one, according to Miss Jenkins's observations:

Mother seemed pleasantly surprised to see Antwone growing. . . . Child has also changed in personality and was able to greet mother with a pleasant smile and permit her to take him on her lap. CW does not believe that child is aware of the relationship with the mother as he treated her as he possibly would treat any other friendly person. This, however, was most gratifying to mother and it was obvious that the visit was the first she has had with the child that was satisfying to her. At no point did child reject mother. CW asked Dwight and Antwone if they wanted to go to the coffee shop for candy. They both said yes and CW observed that mother also was eager to go and expected to be treated like a child along with the two little boys. She chose a frozen sucker and sat at one of the tables with the children who had candy, eating it contentedly.

A week later, Miss Jenkins arrived to pick me up and take me to another meeting with my mother. In those days, I recall, the social workers drove white cars with their government logo on the side. The report described my enjoyment of riding in the car:

> Antwone preened his neck as we drove along the lake shore and looked with interest at the boats, the fishermen and the people. He made no comment. CW did not observe evidence of fear when Antwone entered the building. Mother was awaiting our arrival and CW said to him, "Here's your Mommy, Antwone, say Hi Mommy." Antwone looked at mother intently. He was very sober of expression and there was no indication as to whether he was glad to see her or whether she was just another person he happened to be seeing.

After this meeting, my case was transferred to another caseworker and a lapse of several months occurred before the next visit with my mother. That was in February 1964. According to my file, my mother called and scheduled a visit in March but did not show. Around that time a large gift box of new clothes arrived via social services at the Pickett house. I was told it was from her. In June of that year, she wrote a letter to the caseworker saying that she planned to be married that month, which she believed would make her better suited for my custody. None of the subsequent caseworkers noted what became of those plans. In any event, that letter and a couple of follow-up phone calls a year or two later were her last attempts to contact child welfare in my regard. That February 1964 visit was the last childhood image I have of the woman I was told was my mother. And before long, she had faded almost completely from my memory. For the time being, I accepted as fact that Mizz Pickett was my real mother.

A recurring nightmare from childhood, starting at about age three:

> I'm left in the house alone with Willenda, a neighbor who helps out the Picketts periodically with us. She's in her twenties and doesn't have many friends of her own, girls or boyfriends, and right now she's downstairs in the living room in her regular spot in the armchair in front of the television. She watches soap operas—Secret Storm and The Edge of Night—and smokes Bel Air cigarettes.
>
> Willenda's good-looking, everyone says. Good figure, dark-skinned, with smooth hair and even features.
>
> I'm playing quiet and still as I can, my ears alert to the sounds from outside, hoping/waiting/listening for the sound of someone arriving home, which will make me safe. But instead I hear the dreaded sound of Willenda's voice scratching up the stairs: "Twonny, get down here."
>
> I go slow, still thinking I'll have a reprieve, but she shouts louder, "Hurry up!"
>
> When I get to the bottom of the stairs, she's waiting for me. Doesn't say nothing. Picks me up and carries me through the hall to the door that leads to the steps down to the basement. Carries me down

the steps. Sets me on my feet. For just a second, she just stands there and looks like she's listening for someone coming home upstairs. But hearing nothing, she turns back to me and then—smack!—slaps me to the floor.

Willenda has different voices she talks in and the next one I hear is like a bark, saying, "Now get up. Take off them clothes."

My hands shake as I start to undress and I turn my head 'cause I know she's naked now. She talks in her maddest voice, saying, "I said hurry!" and comes at me, pulling off my pants and shirt, like ripping clothes off a doll.

Shivering, not from cold but from terror, trying to wrap my arms around myself, I close my eyes as Willenda pulls me toward her. Leans down and puts her mouth on mine. Sticks her tongue in. It tastes of the cigarettes she smokes. Her voice changes to soft and low. "Come here," she says, guiding my face between her legs. "Go 'head now, you know what to do." Holds me there, rubbing herself against my mouth, whispering, "Ooooh, yeah. That's sugar, baby."

My eyes are locked shut. But it doesn't help. I know the look on her face while she starts to moan, "Yeah, that's it, right there," moving faster and harder. She has on the worst monster face you ever saw. Only this I didn't make up in my mind. It's too terrible to be true. But it is true.

Then she's finished. Her voice lays empty as she says, "Where your clothes at? Put some clothes on." She sounds like it's my fault I don't have clothes on. She dresses, tosses my clothes at me, and says, "Go on outside in the shade and play. I think they makin' mud pies out there."

For a second Willenda smiles. But then her face flashes a warning. She doesn't even need to tell me in words. I know what it says—never, never, never tell, or something more horrible than you can even imagine will happen to you.

And it wasn't really the fear of her punishing me that kept me from telling anyone all those years. It was the unspeakable shame I felt about what went on with her in the basement, and my unspeakable shame that maybe it was my fault.

A nondescript day from the time before I started kindergarten, about a year after the molestation began, turns vivid in my memory on an early afternoon when all the kids are in school and only Willenda and Mizz Pickett are at home.

Mizz Pickett comes into the living room where I've been watching the noon cartoon show. Now it's going off and she says, "Go on up and take ya nap."

There are strict rules in this house for talking to the woman we know to be our mother. To orders or questions, we must say, "Yes, ma'am" or "No, ma'am," and in addressing her first to ask a question, we call her "Mudeah," a southern black term of respect for a matriarch, a contraction of "mother, dear."

I'm not sleepy, but I've learned better than to say it, so I obey quickly, replying, "Yes, ma'am," going upstairs, and curling up in the army cot I share with Dwight. The whole afternoon passes without me sleeping a wink. Daydreaming as usual, I spend the time cooking up a crazy idea about how exciting it would be if I could go outside and play. I can see me everywhere—jumping up and down the steps in front of the two-story house, running around to the backyard and sneaking into the detached garage back there, and playing with trucks on the driveway. I picture a day with all the grown-ups gone, having a party with my friends from the neighborhood, like Michael from across the street, and giving everybody a tour inside the house. In my mind, they follow me as I take them in the front entrance through the enclosed porch with storm windows all around, into the living room, as they follow me straight from there into the dining room just past an archway. Then, to the right of the dining room, we go into the kitchen and walk around. I show everyone the possums and raccoons in the freezer—like the one I saw one day thawing in the sink. Everybody screams. Me, too.

Of course, in my fantasy I have to let Dwight come, too, 'cause he's the only one who can pick all the locks. Mizz Pickett has everything padlocked—her bedroom, the freezer, cabinets, closets, pantry. She even threatened to padlock the Fridgidaire. But that would've been too inconvenient for Mr. Pickett to get the red Kool-Aid he loves so much. Even so, Mizz Pickett checks the quantity of the Kool-Aid all day long, just to make sure none of us thieving, rotten, hardheaded niggas have taken any.

Between the dining room and kitchen is a little foyer with the lynoleum stairs that go up to the second story, with a landing in between. At the top of the stairs is a hallway. In front of us is a bedroom where one of the Picketts' daughters sleeps. To the right is the bathroom and to our left is the stairway that goes up to the attic room where the Pickett twins and Flo sleep. Down the hall at the front, looking out into the front yard, is me and Dwight's room, beside it Mr. and Mrs. Pickett's.

Then I pretend we go back downstairs and into the

kitchen and down to the basement. Underneath the stairs is Reverend Pickett's locked office where he has his doctoring stuff—specimen jars with nasty things floating in water, scales for weighing the powders and things he grinds up with tools he calls a mortar and pestle, chicken feet hanging on strings, and sparkling crystals for healing ailments. He has other stuff there, too, like foreign stamps and maps.

There's a playroom area with toys, a sofa, and an old television set. In the middle of the basement is a big pole that holds the house up. Just past the playroom is a laundry area with a furnace. Across from it is a crawl space for storing things. Though in my daydream I see us playing in the basement in bright light, it is actually dark and shadowy down there—with only one bare lightbulb hanging from the ceiling.

Finally I hear real footsteps and voices from the living room, sounds that mean the other kids are home from school and I can get up and go play. It's not only Dwight and Flo, but also Mizz Pickett's four grandsons, who come over after school until their mother can get off work to pick them up at the end of the day.

After Dwight comes into the bedroom, hurriedly changing into his play clothes, I hop out of the cot and chase after him down the stairs. Mizz Pickett is waiting at the bottom. Eating possums gives her the power to see through walls. That's how she knows I haven't been asleep. "Twonny," she tells me, "get back up them stairs, boy, and don't come down till ya had ya nap."

Back to the cot I go, only to lie awake listening to the other kids right outside the door in the hallway. Even though I imagine Mizz Pickett will know, I can't keep myself from creeping over and opening the door just a crack to peek out. The boys are in a huddle, one of them holding a book of matches and another saying, "Let's go 'xplore the basement with them matches!"

Another says, "Yeah, yeah, a treasure hunt for them toys Mu-deah hid from you-all, okay!"

Mizz Pickett's powers in force, she yells upstairs to them not to disturb me. I feel more sad now, watching them run off on their adventure before turning to shut my door and go back to the cot.

More sleepless minutes go by. Then I smell something—smoke. And I see it, too, coming from the bedroom heating vent. The idea that where there's smoke there's fire isn't one I understand yet. But the smoke worries me. Maybe I should go out to the top of the stairs and call down to Mizz Pickett, I think. No, I tell myself, she's already yelled at me for not going to sleep; I'll turn over and ignore the smoke. Before long, the room is full of a thick cloud of it and I'm beginning to choke.

Downstairs there are confused sounds and voices yelling all at once. As scary as it is, as much as I try to block it out, the smoke is too much and I get out of bed and go to the door. When I open it, there is fire leaping through the floor and even more smoke. Unable to see the stairs, I go down the hallway to where they're supposed to be and yell out, "Mu-deah, can I come downstairs now?!"

No answer. I yell again. And again. Still no answer. Teetering on the top step, going back and forth about whether to just go down and risk Mizz Pickett's giving me a whuppin', I hear someone calling to me. It's Willenda, saying, "Twonny! Come downstairs!"

"I can't see the stairs!" The hot flames are jumping and grabbing all around me now. The smoke has turned as dark as tar, and I can't breathe. On my knees at the top of the stairs, I suddenly see and hear Willenda through the smoke reaching out to me. She picks me up and carries me down the stairs, through the house, toward the front door—all amid the smoke and fire—until we are finally outside, where I join the rest of the entire household standing across the street with neighbors watching the house burn.

The fire trucks come, and here the memory I have of the day I nearly died in the fire fades to black.

Months have passed since the boys burned the house down with matches. It has been repaired and remodeled. This day, the kids are at school, Mizz Pickett is at her job as a domestic over on the west side of town, and I'm left alone with Willenda. She's watching *Secret Storm,* smoking her cigarettes, lost in the lives of these white people on TV. Being a master of watchfulness around her now, I know she usually won't take her eyes off me. But as long as the show is on, I'm free to escape her sight. And that's what I do.

In the kitchen I find a pot simmering on the stove. I don't want to check its contents, just in case it's some wild animal that I've seen in Mizz Pickett's pots before, with tails and claws and eyes boiling away. It's the blue flame underneath the pot that entreats me to look closer. After staring at it for a bit, I go over to the corner where the straw broom is kept and pull out a long single straw. Back at the stove, I push the straw through the fire, watching it melt away. I do this several more times, feeling quietly powerful, until the final time when I turn and feel my chest contract and my heart jump into my throat as I see Willenda standing right over me.

She smacks me hard and sends me upstairs to wait for Mizz Pickett to return home. The pain in my chest continues—a sharp, knifelike stab that makes my breathing shallow and difficult. It is a chronic condition that will afflict me for years.

In the evening, I'm still waiting in the bedroom when I hear Mizz Pickett's voice calling loud, "Nigga!" By this time we could discern which of us she meant—Dwight, Flo, or me—by the tone of voice as she said the word *nigga*. That was the brand of disgust she used for me, so I go down to the kitchen, where I see her standing over the stove lighting a rolled newspaper. Willenda's at Mizz Pickett's side, her head cocked sideways, her expression smug.

"Com'ere," says Mizz Pickett, and I walk up to her, trying to wear my most innocent smile. She just smirks, shakes her head, and asks, "So, nigga . . . ya liiik' fiah?"

"No, ma'am."

Mizz Pickett grabs my arm and begins to beat me with the flaming newspaper. I scream and scream as she and Willenda laugh and laugh.

For a long time we never heard anything about the Picketts being foster parents and there was no question but that they were our mother and father. Just as we called her Mu-deah, the Reverend Pickett we called Dada.

My first memory of him is in sepia tone with only a little color. Like you might see in old-fashioned photographs hand-painted with pale, Easter colors—lavender, peach, light yellow. Maybe it's because I had the feeling he liked me, even though he barely said anything to me, but I liked him. And maybe that made him that small amount of color dabbed into the otherwise darkness that was my childhood; or maybe it's because the first image I have of him comes, in fact, at Easter time.

Mr. Pickett, a strong, stocky, dark-skinned man, whose nose is broad and whose head is bald, is standing at the front door of his storefront church on St. Clair, unlocking it. He is wearing one of his many suits. Khakis and suits are all he ever wears—khakis for going to work cutting grass and suits the rest of the time. He sends his shirts to Swift Cleaners, on St. Clair, too, where their logo is a man in a bellhop uniform running with streaks of speed coming off of him as he delivers a freshly laundered shirt. Swift Cleaners makes Mr. Pickett's shirts a marvel of pressed starch.

He was the kind of man, Reverend Pickett, who said a lot without words. Sometimes he'd walk past me at the dinner table and grab my head, a nubby football, and rub it just to show he noticed me there. From time to time, when no one was looking, he'd give me a nickel. I'd look for a smile, but there was never one. That's okay, the attention felt good.

One of the reasons I think he liked me was that he got to practice his doctoring on me. At Pickett family gatherings, Mizz Pickett would tell the story of the time when I was two and had eczema from head to toe. None of the doctors referred by the social services office had been successful in treating my ailment, even by restricting my diet from the foods they thought were causing the allergy. "Well, suh," she would tell the story, "the young'un was sufferin' and Day-di"—she called him that affectionately, making it rhyme with *Sadie*—"he made a salve for the rash that cleared it up real good inside a month. Let him eat all the eggs, oranges, and tomatoes he wanted."

And Mr. Pickett would look on quietly with pride.

As she bragged on her husband's success with me, I could bring into my senses the feel and smell of the salve that was layered on my whole body. It was as grainy as fine sand, with the pungent odor of sulfur, and burned like salt on a raw wound. That salve was more suffering than the rash, just about. But it worked.

Watching Mr. Pickett unlock the front door of church this Sunday, I feel like he values me, maybe 'cause I'm proof that he's a real healer, a minister of God and medicine.

This is one of a thousand Sunday mornings that begin at the Pickett house like holiday mornings full of excitement and importance, as everyone rushes around preparing for church. There are three of us younger kids to get ready, along with Elaine and the Pickett twin teenage girls, Mercy and Lizzie. For twins they don't seem anything alike to me. Mercy is nice and nice to look at, in a simple way; Lizzie acts and looks mad most of the time. Sometimes, like today, the Picketts' other older daughters and sons come to the house and attend church.

In church, where the turnout is good this Sunday, there is an air of added excitement, a nervous energy I can feel but don't understand. At testimony time, the first woman has a story about how she had tonsillitis and yesterday morning she prayed for five hours straight and the Lord healed her before she'd even finished praying: " 'Cause de Lawd knows ya desire, before ya even ask him. And He'll grant ya desire, if ya only ask."

Over the course of the next fifteen minutes (which feel like an eternity), the stories of the other testifiers become more and more fantastic—more healin's; savings of mortgages; food miraculously on doorsteps; and crooks, liars, and thieves all brought unto Jesus—until the Lord Himself arrived in a beam of light in the bedroom of one woman. Finally, the church, unable to take any more, breaks into a spirited rendition of "The Old Landmark."

Then Reverend Pickett is at the pulpit, looking like the

sheriff, talking loud and fast, running the Devil out of town. People are in a frenzy, dancing, clapping, stomping their feet, fanning themselves, breathing in and out with deep sighs, even crying and talking in tongues and shouting things like "Praise the Lawd!" "Glory!" and "Hallelujah!" pronouncing this last with different foreign-sounding accents (French, Spanish, and Transylvanian)—Hal-*le*-lujah; Halle-*lu*-jah; and *Holi*-lujah—each trying to sound more godly than the other.

There is a lady in front of the altar, jumping around and screaming like crazy. It's what they call catching the Holy Ghost. The lady looks like she's in a lot of pain, but everyone is on their feet, cheering her on, like they want that Ghost to drive her mad. Maybe the Holy Ghost is friendly, like Casper on television, 'cause the grown-ups seem to like him.

But I don't know, he seems kinda mean to be making that lady act that way—falling on the floor and scooting around and crying. Now I wish he'd let her go. The two deacons—sorta like the deputies of the church—probably are wishing the same thing as they come running down front.

They unfold a large white sheet and lay it over the lady so nobody can see her panties under her dress. But it's too late, they're pink.

My eyes are too glued to all the excitement to turn and see Dwight's reaction. The focus of my attention is the man at the center of it all, Reverend Pickett. He is standing behind the podium in the pulpit, happy as can be, clapping and stomping around like this is the very day of jubilee—the most powerful man in the world to me.

Looking up to Heaven, beaming, he doesn't even notice that the lady is all tangled up now in the sheet. I shake my head, finally glancing over at Dwight and Flo, who are probably thinking what I am—that lady's gonna make us be here even longer today and then they might forget to give us cookies after church.

For the first six years of my life I waited on an island of uncertainty, confused about how I'd become shipwrecked there in the first place. Then, on the day after Halloween 1965, came the equivalent of a message in a bottle that washed up on my shore.

It's the morning after trick-or-treating, and our shopping bags are filled with candy. Mizz Pickett, dressed in her everyday housedress and slippers with her stockings rolled to the knee, her glasses as usual sliding down her nose, sits in the living room going through each bag. She is serious and thorough, tossing some of the candy into the wastebasket and some into a metal box she always keeps under lock and key, like everything else. We're quietly seated in front of her—me, Dwight, and Flo—each of us waiting for her to give us a piece of candy. Since we weren't allowed to have any before going to bed, I have spent the night in candy dreams, almost tasting the sugar in my sleep.

It was the first Halloween I got to go out trick-or-treating with just kids and, lucky me, the neighbors had been really generous. It was fun to dress up and pretend to be somebody else—a skeleton in my black nylon one-piece with white bones on it and a skull mask. But I mostly liked looking at the other kids in their costumes and seeing the houses all lit up with carved pumpkins and handmade decorations.

As a black working-class area of Cleveland, our neighborhood of Glenville was well kept in those days. Being on the edge that bordered the industries close to the lake, our block wasn't the fanciest. Often there was a metallic smell in the air, heavy with the odor of oils used to keep machinery in motion, mixed with the scent of fish from Lake Erie. Sounds could be heard day and night from the factories nearby, along with noises of trucks grinding their gears in the gravel lot across the street. At 1 A.M. every morning, a train passed just behind the Picketts' house, rattling the whole block. In spite of all of this, everyone around us worked hard to make their homes welcoming, making the most of what they had.

Everything looked nicer to me in the fall, my favorite time of year, when the leaves began to turn those beautiful burnt-orange fall colors. I loved the smells of autumn and the crunching of the leaves under my feet as I walked to and from school. Even though winter was on its way, this time of year reminded me that there would be a new beginning after that, a chance to start again in spring.

Waiting for Mizz Pickett to finish her sorting, I relive the highlights of the season and the night before as my mouth drips in anticipation of its first sweet treat. Finally, she has finished sorting every last lollipop, peppermint, chocolate kiss, Hershey's bar, gumdrop, licorice stick, piece of bubble gum, PayDay, Jujubes, you name it. Then, wiping her hands together first, Mizz Pickett slams the lid shut on the metal box and slaps on a Master lock. (She swears by them.)

"Now," she says.

Here it hits me—she ain't gonna let us have none.

Sure enough, Mizz Pickett stands up, the metal box under her arm, and says, "I'll give ya some uh dis here candy when you deserve some uh dis here candy . . . and not until. Now get on out from under me."

We drag ourselves and our aching sweet-tooths up the lynoleum stairs to the bedroom I share with Dwight. The taste of candy we almost had turns sour in our mouths as we try to form words we've never said out loud about Mu-deah. Her last statement still burns in my ears: "when you *deserve* . . ." It's not fair, and I know it.

Too upset to sit down, we assemble by the bedroom window. Outside the leaves on the trees shine golden in the autumn morning sun, but it doesn't make me feel any better. The bedroom is small and bare except for a bed and a cot and a dresser. Dwight leans against the wall with arms folded. He is a scrappy seven-year-old, thin and outwardly agitated most of the time. A good-looking kid, Dwight has smooth, tea-colored skin and even features. Always on guard, he glances at the door and finally breaks the loud silence by saying in an angry whisper, "Ain't nobody else's mother locks their trick-or-treat candy in no box!"

"Yeah," I echo him, "ain't nobody else's mother mean like her, neither."

"Yeah," Dwight echoes me, "how come she gotta be so mean?"

Flo's been only listening until now, looking at us like we're a pair of slaves, six and seven years old, just fed up and planning a revolt. She is an average-looking ten-year-old girl who tries to blend in—except for her hair, which she wears in a style she calls The Florence, named for herself. It's parted on the right side and at the back of the crown, like an L, and her hair is curled and dressed with Crown Royal hair grease, then combed smooth.

Flo gets this solemn look on her face as she explains carefully, "She's not our mother. We got another mother, and it's not her."

"Who is our mother?" I ask in a hushed voice. In the back of my brain I feel the stab of a long-lost memory of a woman I was once told was my mother. It still isn't clear. I wait for Flo to explain.

She shrugs. "I don't know who your mother is, but me and Dwight got the same mother. We're all foster."

Foster. The word flies up and smacks me in the stomach. Only I don't know what it is to know if it hurts or tickles. Dwight, though, he's got a bad feeling about the word. I can tell by the look on his face, like he just smelled something that stinks. "What's that?" he asks Flo.

"It's when you gotta live with somebody who ain't your mother, that's what."

"Oh," he says, coming over to lean on the window ledge by me. "Well, I don't want her to be my mother anyway."

"Me, neither," I say.

Dwight scrutinizes me and then Flo, asking her, "If you and me are real brother and sister, why your last name is Hill and my name is Perry?"

"Maybe we ain't got the same daddy."

"Oh, well, I'm glad she ain't my mother."

"Me, too," I say again. I look over at Flo, asking her, "What about you? You glad she ain't your mother, too?"

Flo sits down on the floor, thinking about it for a second as she curls up on her side. Her answer: "She's better than having a mother that don't want you."

"You think she wants you?" Dwight shoots at Flo, crouching down on the floor, too.

Flo doesn't answer him. But she does say that Mizz Pickett gets money from social services for keeping us.

I stand at the window thinking by myself, starting to get worried. For all this time I'd concluded that Mizz Pickett was my mother. Now I hear she ain't my mother, and Dwight and Flo's mother ain't my mother. Then who is my mother?

That question is one that will go unanswered for the following ten years, when a second message in a bottle will offer me my next clue. In the meantime, Flo's news this morning, shocking and worrying as it is, also fills me with relief, as if a weight has been lifted. With a sigh, I plop down on the floor beside Dwight and Flo, feeling good to know for certain that Mizz Pickett is not my mother. Even if I don't know who my real mother is, I'm sure now that she's a young, pretty, nice lady who would let me have a piece of my own Halloween candy if I wanted it.

We all sit quiet for a long while, just thinking.

And I believe we fell asleep together, right there on the floor.

A recurring dream from childhood:

I'm standing alone in a field, knee-high blades of grass blowing gently around me. I have on this old-time white cotton nightgown. It's clean and sweet smelling and really bright white, almost holy, like I'm dressed for an important occasion. But all I see before me is an ocean of whirling shiny green grass and all I hear is the low purr of the wind.

I can feel the warmth of the noonday sun shining behind me and I reach out my arms and tilt my head back, breathing it all in—a deep, deep breath. And there above me is a rainbow lacing across the sky, a sky colored by multiple shades of blue. Pale sky blue, dark royal blue. Blues I never even heard the names of: cerulean, aqua, cobalt, azure, indigo.

Then I turn to look behind me, toward the sun, and I see this full-figured, middle-aged woman with dark brown skin my color and with a wonderful smile. She has on a burlap dress and a white cotton apron and a white scarf tied on her head. She looks at me, smiles even wider, and takes my hand, lead-

ing me through the grass until we come to a large, dilapidated old barn. Through the splintered wood, I can hear wondrous voices inside and the sound of African drums beating in celebration.

Suddenly, the big doors swing open and a tall, muscular, barefoot blue-black man steps into the doorway. He's wearing a white cotton sleeveless shirt with burlap pants, cut off and frayed at the knees. Leaning backward with his fists on his waist, he looks down at me. And I look up at him and see right into the wide nostrils of this glorious man.

He looks so serious until his voice comes out, deep and loud, joyful, "Ha, ha, ha, haaaa." He slaps his chest and extends his arms. I understand he means to say: "Ya home now, boy, we been waitin'."

The man takes my other hand and together the three of us, the woman and him, me in the middle, we walk into the barn. There are two levels of rafters on each side of the barn; they're filled with people, all ages, all of them dressed in the same clothes as the man and the woman. In the middle of the dusty, dirty floor is the longest picnic table I ever saw, covered by a red and a white checkered tablecloth. At the far end of the barn, I see the musicians playing.

The music and the voices and the laughter are splendorous.

Instantly, the people on the rafters empty onto the main floor, smiling, nodding, and whispering elatedly to one another. Their whispers seem to be saying, "It's him, he's here," words I can't hear but can only feel in this syncopated song of welcome.

Everyone follows as the muscular man hoists me

into his arms and sits me at the head of the table. The woman leaves for a moment and before I know it, she's back, with two plates, each holding a tall stack of pancakes—one for me and the other for the man. After she sets them down, he takes a pitcher of maple syrup and pours it over my pancakes for me. Now everyone else sits down all around the table, with their own pancakes, watching, grinning, waiting for me to start, and I feel a familiarity about them all. Like I know them, too.

With a fork and knife, I cut a big piece, stir it around in the syrup, and finally taste the most delicious treat I've ever had. Now we're all eating, enjoying every last bite. Before I know it, the pancakes are all eaten, and it's time for me to leave.

The entire gathering rises to escort me to the barn doorway. They say their good-byes and stand back, the proud man just in front of them. I start to step outside, back into the field and the world I came from, where my path lies ahead strewn by shadows. I turn back into the barn, not wanting to leave. It is then that the woman with the wonderful smile comes to me, puts her hand on my shoulder and leans down close to my ear, speaking words I don't recognize but that sound familiar, as if from an old song I'd heard a long, long time ago.

I always woke up from the dream with the memory of her smile and the feeling that I could survive the coming day.

It was such a great dream, something that belonged to me and to nobody else. When it was bedtime, I'd even try to concentrate on the images—as if I could turn the dream channel in my mind like the dial on the television and have the dream that night. But it never worked that way. In fact, it seemed that in sleep the dream never came when I wanted it, only when I truly needed it.

In time, I began to understand what the woman was saying to me at the end of the dream. It was something like this: "Be mindful, chile, and look out fo' yo'se'f. Don't worry none, you're not alone." Against all the realities of not knowing who I was, where I came from, or where my place was in the world, this led me to believe that I was somebody and I belonged somewhere.

They may have just been words from shadows in my dreams, but I took them to heart. They became a lighthouse, distant yet visible, in the tempest of my youth, guidance I would surely need in the greater storms to come.

two

I started my school career with great promise, according to my state child welfare report:

> Antwone was registered at Oliver Wendell Holmes School in 9-64. His adjustment has been good. Grades received for semester ending 1/65: Conduct B; Effort B; Appearance A; "Antwone is a nice little helper and tries very hard to follow directions." Grades received for semester ending 6/65: Kindergarten A; "Antwone is a good citizen."

At first, going to school brought with it a new freedom—temporary liberation from the Picketts—and I found that I loved to learn. But this was short-lived. Soon the problems in my home life began to travel with me to school. By the first grade, I was no longer seen as a child with promise but as a painfully shy little boy who walked with his head down, stumbled on words when called upon, and had to be reprimanded for daydreaming in class.

All of this went unknown to the various social workers assigned to my case during the years that I was five, six, and seven years old. In fact, they noted that I was doing well in the Pickett home. Mizz Pickett had convinced them regular visits weren't warranted:

Called the foster mother to introduce myself. She said they were having no problem at all with Antwone
and that she felt she could get along
on a minimal contact basis. I did
tell her that I was planning to make
an appointment to contact her but
have since written to say I am leaving the agency.

For three years, she kept the social workers at a distance. With heavy caseloads and far too many serious problems being reported, they were probably very relieved that Mizz Pickett had no complaints, and they didn't question her further.

Meanwhile, things took a turn for the worse after Flo's revelation that we were foster kids. Mizz Pickett was livid and whipped Flo till the sun went down. With Dwight and me, whose crime was knowledge, our punishment was more psychological. Mizz Pickett took every opportunity to make sure we didn't get uppity ideas. Almost every day she'd warn us, "I'm gonna take you rotten niggas back where I gotcha."

I assumed she meant the hospital. That's where I thought all babies came from. But Mizz Pickett had a speech ready just for me. "When they broughtcha to me you were two years old, covered from head to toe with sores, and I nursed ya back. And they tole me you were retarded and I beat the sense into ya." If I opened my mouth to say anything, I would get part two—the LP version of the speech. When she was in her fussing mode, Mizz Pickett put her left hand on her left hip, jutting it out, and leaned her upper torso into me; with her head rocking from side to side, she shook her right index finger at me almost at the tip of my nose. "And soon as ya get old enough to piss over a boot, you wanna sass me?" Now we come to the chorus: "Nigga, I'll take ya back where I gotcha from."

Just in case any of us were harboring hopes of having our real parents come for us wherever it was she got us, she let us know again: "You-all wit' ya no-account mammies, they don't wont-cha. Don't nobody wont-cha. If they did, ya wouldn't be here. But don't think I wont-cha. I jus' gotta pay mah notes." The shame of parents who didn't have bank accounts was confusing enough, but why anybody had to pay money for notes I really didn't understand.

Finally Dwight and I came up with a desperate measure. We decided that we were going to report Mizz Pickett to our caseworkers. What a heady day. It was dangerous, we knew, but thrilling at the same time to think we weren't completely powerless. Then Dwight blurted it out one night after she gave us both one of those good whuppin's. He stood up and said defiantly, "I'm gonna tell my social worker!"

Mizz Pickett stopped dead in her tracks and started laughing. "Ha! Social worker?" she scoffed. "Nigga, I'm ya social worker. I'm them white folks' boss." Still laughing, she left the room.

Dwight and I didn't say a word. She shut me down good, the same as if she had beat me unconscious—something she would do in time to come.

All hope of rescue gone, from then on I resigned myself to living in a combat zone. This meant learning to be on constant high alert, reacting to most situations either with flight or fight, and seeing others as either friend or foe. Still, the main enemy knew well how to divide and conquer, and she saw to it that there were no further plots of insurrection.

Dwight is laughing, sitting on the edge of the sink with Reverend Pickett's shaving mug in one hand and the brush in the other as he covers his mouth, chin, and cheeks with shaving lather.

Me, I'm a seven-year-old mad scientist, mixing a concoction in the sink of mouthwash, iodine, witch hazel, and a few other items in the medicine chest. Strangely enough, Mizz Pickett doesn't keep it padlocked like everything else in the house.

It's after church, a little past 8 P.M. on Sunday. Church can be every evening on weekdays and on Sundays, it's all day. We sat through three services, clapped and sang and did everything we were supposed to. Now Flo and Mizz Pickett are downstairs, the rest of the Picketts haven't gotten back from church yet, and, instead of washing up for bed, this is one of those times when a crazy fog has come over us, and we're acting like normal, mischievous children.

Dwight and I giggle at his face in the mirror and my mess in the sink. We're so delighted we don't see the approach of Mizz Pickett. Suddenly she's in the bathroom, and we're in trouble.

Down to the basement we go, where she ties us to the big pole that stands from floor to ceiling—a stanchion, they call it, because it holds the house up. She ties us to it back-to-back with rope and leaves us here in the dark.

After a few minutes, Dwight whispers, "Don't be scared."

I say nothing, 'cause I am scared. There's monsters and bad things happen down here. Dwight and I know we'll be in more trouble if she hears us talking, so we sit there quiet, except for our short, frightened heartbeats and the sounds of unseen creatures in the dark.

Mizz Pickett ties us up here a lot, before or after whippings. We've been left here for better parts of days. I stare at something I can identify—the water heater with its tiny red light that blinks off and on, a dragon's eye standing guard over us—and try to let my mind wander.

My current favorite daydream is about a little girl in my second grade class. She's a teacher's pet, and she has this pretty yellow dress and wears perfume. Thinking about the girl in the yellow dress, I can bring her pretty smell into the darkness and make it hug all over me. I close my eyes—like in the song by Peaches & Herb—and take a deep breath. I see myself walking and talking with her, saying how cute she is and stopping by the lakeshore to hand her a flower I've picked myself.

In reality, the girl in the yellow dress floats past me where I sit in the last row of class on her way to the front row, never even noticing me. But in my mind's story, we love each other. To me, she's "My Girl" in the Temptations song—sunshine on a cloudy day.

All of a sudden, the light over the stairway goes on and I become more scared, realizing it's Mizz Pickett on her way down to beat us. But as her footsteps start down the stairs, I hear that they're not heavy enough to be Mizz Pickett. She weighs two thousand tons, so it can't be her.

The overhead light comes on as Flo steps off the last stair. She's got two mounds of ice cream in one of Mizz Pickett's good bowls. Matter-of-fact, Flo sits down on the floor, her back to the wall, eating the ice cream, licking the spoon, smacking her lips.

What a traitor, gone to the other side. And for what price? A bowl of vanilla ice cream. Mizz Pickett probably sent her down on purpose to torture us with the ice cream.

She just sits there, not saying a word. Then Dwight can't take it anymore and begs, "Let me have some, Flo?"

"Nope," she replies. "Mu-deah says you-all can't have none."

I don't care, I think, chocolate ice cream's what I like anyway.

Flo takes the last two bites, licks her spoon one more time, and lets it clink down in the empty bowl. Then she adds, like she almost forgot, "Mu-deah said when Dada gets home, she gonna make him give you-all a good whuppin'."

Right on time, I hear Mr. Pickett's station wagon pulling into the driveway and my heart starts beating real hard.

Flo jumps to her feet and runs upstairs, turning the lights back off as she goes. We wait in the dark again, this time both of us beginning to shake.

When the lights come on next, it's Mr. Pickett. I hear him coming before he appears and I know it's him 'cause he weighs a million thousand tons.

Reverend Pickett steps off the last stair, holding his gigantic leather belt folded once in half. He pauses there at the bottom of the stairs for a moment, looking at us like he can't believe Mizz Pickett tied us up.

Dwight seizes that moment and begins to try to talk his way out of the whuppin', starting, "Dada . . ." but loses steam as soon as he sees Mr. Pickett coming over. He unties us and unfurls his belt, whipping us both.

Sometimes I think Reverend Pickett only gives us whuppin's because Mizz Pickett nags him so. Usually they're halfhearted and it seems he's been whipping us less and less, although this night he's clearly serious. I get off easier than Dwight, who runs from Mr. Pickett, causing him to chase Dwight around the basement and forget to come back for me. More and more, Reverend Pickett seems to favor me over Dwight. This only creates further conflict between my foster brother and me, the seeds of which Mizz Pickett is already busy sowing.

A month before my eighth birthday, in July 1967, after a three-year lapse without any of the social workers noting anything of significance, my case was transferred to a Miss Vicki Blum, who now reported many of Mizz Pickett's concerns about me.

One of the problems was that over the past year Dwight and I had both begun wetting the bed—for which (the caseworker never knew) Mizz Pickett threatened to cut off our penises. She called them "ya little men."

Since Dwight wet the bed more often than I did, he started getting in trouble with her more often than I did. He reacted to that by teasing and picking on me. My being younger and maybe treated better than him by some in the house made it worse. It wasn't only Mr. Pickett that seemed to favor me, but also Mercy—the sweet, good-looking one of the twins.

Before she'd go in for her bath at night, she used to find me and say, "Twonny, clean the tub out after I take a bath and I'll give you a dime."

I'd sit outside the bathroom door, excited as I waited for her to finish so I could earn my ten cents. It wasn't just the money or the candy I bought with it that filled my pockets, it was Mercy's kindness that lit up my days like a black-and-white movie turning into Technicolor. They were small gestures—a soft look, a warm pat on my shoulder, a gentle voice—but I counted and collected them and never forgot them.

The nicer Mercy was to me, the more hateful her twin Lizzie became. Once she gave me two dimes and promised, "Go to the store and get me a bag of potato chips and I'll give you ice cream from the freezer." Being little then, I knew it would take me five times as long to go to the store and back as it would for her to get up from her chair in front of the television and go to the kitchen and back. I knew how tired I'd be. But all I could think about was getting the ice cream, so I was willing.

When I returned, proud of myself, and handed her the potato chips, Lizzie tore open the bag and started eating them, waving me away. I reminded her about the ice cream, but in between crunching on a full mouth of chips, she said, "Later." That was her answer every time I asked over the course of the rest of that day. It was only when I climbed into bed in the night that I realized Lizzie never intended to give me ice cream.

I didn't cry about it then, but my feelings were so hurt they remain so to this day.

Mercy seemed to notice when her mother or Lizzie was being mean. Maybe that's why she took extra effort to make me feel better. But since no one paid special attention to Dwight, Mercy's kindness to me agitated him more. So he took it out on me. We began to fight constantly.

We woke one Tuesday morning after Dwight wet the bed. In the past, he'd tried turning the sheet around so it appeared that I wet the bed. But since his pajamas always gave him away, he took to hiding them. This morning, he pulled off his pajamas and hid them in the vacuum cleaner, got dressed for school and forgot all about the bed-wetting. Mizz Pickett didn't detect anything about the bed that morning, either.

But on the following Saturday, the day set aside for major housecleaning, Mizz Pickett tried to use the vacuum but the suction was weak. She opened it and, as she later described it, the smell knocked her to her knees, causing her to have what she called a dizzy spell.

When she came to, she yelled out, "Niggas!" Dwight and I stopped our house chores and ran to her side.

"Which one of you low-down niggas put-cha pissy nightclothes in this here vacuum?!"

"I don't know," we replied in unison, shrugging our shoulders.

Mizz Pickett left the room, only to return shortly with an extension cord, promising, "I know how to get the right one."

This meant that she was going to whip us both. As she raised the cord, I quickly stooped and pulled the stiff pajamas out of the vacuum cleaner, held them high and said, "These ain't mine." And Dwight didn't wait for her reaction. He just took off and she took off after him, swinging the cord and yelling, "Nigga, don't you run from me!"

On another occasion Dwight had run from her and made her chase him all around the house, downstairs, around the living room, down into the basement, and back upstairs again. By the time Mizz Pickett caught Dwight, she was too tired to whip him. Sweaty and trying to catch her breath, she just gave him a shake and let him go, then stumbled into her bedroom. A little while later, we heard her snoring like an old bear in winter. Dwight and I laughed about it for months.

After that, she started locking doors behind her before she came into the room to give us a whuppin'.

In July 1967, when Miss Blum was transferred to my case, Mizz Pickett had a new complaint about us. She told my caseworker that although the bed-wetting was no longer a problem,

> . . . child and brother like to uri-
> nate in waste baskets, closets and
> bottles during the day.

This was false. Why Mizz Pickett made it up, I couldn't say. What I later thought was she had decided that if the state knew what a difficult ward I was, it would give her more money to keep me. But that was not the state's policy and, contrary to Mizz Pickett's desires, Miss Blum became very involved in my case. For starters, she insisted on an overdue physical exam, which resulted in:

> Recommended circumcision on elective
> admission basis. . . . He was admit-
> ted to University Hospital on 8/6/67
> and operated on 8/8/67 and was dis-
> charged on 8/9/67. He was well pre-
> pared for this surgery, had support
> of foster family who visited fre-
> quently.

I don't remember either of my foster parents or my foster siblings visiting.

Later, when I received a visit from my social worker, she asked, "Antwone, is it true you told the nurses that you didn't want to go home?"

"Yes, ma'am," I answered. "I like it here."

"Why?"

"Everybody's nice here."

This gave Miss Blum pause. She asked, "Aren't they nice at home?"

I didn't answer.

My silence wasn't enough to prolong my hospital stay, so I returned, as planned, to the Pickett house, where I was soon replaced as the youngest in the home by a new ward. Keith was four, with big sandy curls and light tan skin. Mizz Pickett treated him like visiting royalty. He slept in our room in a twin bed by himself, while Dwight and I stayed in the smaller cot together.

In this world where before I thought everyone was black—except for the white people on television and the ones at the hospital and the welfare office—Keith now represented something in between. And the message I got from his special treatment was that being lighter-skinned made him better.

It was also around this time that Lizzie, the mean twin, began to tease me by calling me "Chocolate" and other names that had to do with my color. Whenever she talked to me her voice was full of resentment, but when she called me Chocolate or Hershey's or $100,000 Grand Bar, it dripped of awful loathing mixed with delight that I was so low as to be nothing. It made me feel that being dark as chocolate was the worst disgrace, something I couldn't do anything to change. And I started to hate my dark skin.

The attitudes at school didn't help. Although it was 1967 and civil rights were on the march around the country, Cleveland remained strongly segregated. Like the other schools in the Glenville area, there were only black students and mostly black teachers at Oliver Wendell Holmes. It was my experience that the lighter-skinned children with the whiter features and less coarse hair, like the little girl in the yellow dress, were more liked by the other children and even by the teachers. They were the ones who always got to stay after class and clap the erasers, something I was never chosen to do.

The time I remembered being singled out was when a teacher stood over me with her finger pointing down at my hair. She was half laughing, half scolding me for the zigzag line that was my attempt to part my hair. Maybe not everyone in the class turned back to me in the last row, laughing uproariously as they pointed, too, but that's how it felt in my memory.

By the third grade, school was becoming almost as much of a combat zone as the Pickett home.

This fall there is an Indian summer, when the days remain warm and dry longer into the year than expected. In the past, this small breath between the seasons has always made me happy, but this year it does little to ease the heaviness I feel. At a certain point, trudging to and from school, I start to notice that I'm changing. I can't seem to go to that place in my mind where I used to find happiness.

That was a time when I was younger and I became aware of how the morning sun came through my bedroom window, making a shadow on the wall of the entire windowpane. Every morning I watched the shadow and tried to catch it as it moved across the wall, but I never could. But when I left the room for a while and then came back, the shadows would have moved in my absence. It was very entertaining.

Downstairs in the evening was another show when the setting sun moved to the other side of the Pickett house, its golden color brightening the whole dining room. Many evenings I sat beneath the huge cherry-wood dining table till the sun gave way to the pale light of the moon. In that quiet hollow, I felt safe and good, imagining that the sun planned its day of activity just for me.

But now I've lost my ability to imagine a relationship with the sun, the shadows and the golden light. Other magical sights have stopped captivating me—like the two-story buildings at the intersection of St. Clair and 105 where Cleveland Trust Bank was on one corner and on the other corner was a Rexall's with an escalator; and farther down the same street where we drove past in the car at night and I was dazzled by the brightly colored neon signs flashing martini glasses and the word LOUNGE. I could no longer see the magic.

This was when I changed, when I had to divert and channel my powers of imagination to my defense against the increasing difficulties surrounding me. My daydreams were no longer wonderful adventure stories that took me away from the troubles; they were just distractions that gave me a few minutes of rest. So I stopped believing in them and my stolen pleasures came to an end. I stopped melting snow and started resenting winter. And I stopped seeing the Picketts as alien monsters and started to see them as mean people.

Every morning Dwight and I walked up 105 among the many groups of boys and girls on our way to school and came to the candy store. Every day, we followed the other kids as they stopped inside, the door opening and closing with a tinkling bell, as the owner, Mr. Murphy, called out a cheerful good morning.

He was friendly-faced, short, and pudgy, like Spanky from *Our Gang,* and the only white man I ever noticed in the neighborhood. Possibly, he was a leftover from the days when this all-black community used to be white and Jewish. There were a few other reminders of that time, like the many Stars of David that remained etched into most of the bigger and older churches. In fact, it was two Jewish high school students from Glenville High who invented Superman.

Whether or not Mr. Murphy was Jewish himself, I don't know, but his store had been there since those earlier days. It was a treasure chest inside, with a magnificent array of different kinds of candy, everything sparkling in jewel tones and shiny, brightly colored wrappers. The other kids took turns picking out their treat for the day, each with their pennies and nickels ready. Day after day, this was the routine. But nothing for Dwight and me. Since we weren't allowed to have any money at all, we could only stand and watch, swallowing our regret in our own different ways.

One morning when nobody was upstairs with me, I went into Mr. and Mizz Pickett's bedroom and took two nickels, hid them in my shoe, and went off to school. Even though Dwight and I usually walked together, I was able to lag behind enough that he didn't see me when I stopped in to Mr. Murphy's store. For the first time, I felt the thrill of picking out a handful of two-for-one-cent candies, paying for them, and having money left over. On my way to school, I felt an even bigger thrill of sharing my candy and my leftover pennies with the other kids.

The next morning, I took money again—a nickel or two, and a dime. The following day, I did the same. This continued for some days until, without warning, it was announced in class that I was wanted in the principal's office.

That was how I found myself, scared and guilty, standing in front of Mrs. James, our stern, dedicated elementary school principal, having to answer to the reports that I'd been giving money away to the entire student body. I knew the rumor had run amok, but I also knew better than to contradict her. What's more, I knew to tell the truth when she asked where I got the money.

"Off Mizz Pickett's dresser. Sometimes I get it out of her pocketbook."

Mrs. James nodded, walked to her chair behind the desk, sat, and gestured for me to sit down in the chair across from her. After a long silence, she asked, "Why, Antwone? Why did you take money that doesn't belong to you?"

"B-b-b-because," I heard myself stutter, "if I give the kids candy and money, they like me and want to be my friend." I paused and then admitted, "Nobody liked me before."

Mrs. James said, "Good boys don't steal, Antwone. You understand, don't you, that I have to call Mrs. Pickett and tell her what you've been doing?" She posed it as a question, but it was a statement of fact.

Upon hearing that, my heart felt ready to pound out of my chest. There was the sharp pain again that restricted my breathing as I lowered my head and murmured, "Okay."

Mrs. James led me to her outer office, where I sat in one of the smaller chairs provided for the little children and watched her return to her office, closing the door behind her.

Mizz Pickett arrived in no time and from the moment she looked down at me in the small chair, I knew I was going to get it. My heart was beating so fast that the sound of the blood flowing inside my little body drowned out everything else, and I didn't hear a word that she and Mrs. James discussed about the situation.

The next thing I remember was being in my bedroom naked, tied by my arms to the end of the cot, and she was whipping me with a switch made from a thin green branch from a bush in the backyard. The whipping continued until welts popped up on my skin and began to open. This was the day Mizz Pickett would brag about for years to come, the day when I was eight years old that she beat me unconscious.

It turned out that Dwight and Flo had been taking money from Mizz Pickett, also. No haul among the three of us ever garnered more than sixty cents. But every morning from then on, we'd submit to a strip search, with threats that we'd all have to endure a cavity search, too.

These misdeeds of ours made for long, elaborate retellings by Mizz Pickett. Later on, I came to recognize that she wasn't a bad storyteller, in spite of her exaggerations and fabrications. I also came to think that Mizz Pickett relished the idea of being victimized by us, as if by telling everyone how difficult we were would show how charitable she was and what a sacrifice she was making by having us. In the evenings, she used to cozy up with the telephone in the dining room and talk for hours to anyone who would listen about how terrible we were. Many nights, I'd sit on the linoleum stairs listening to her start every conversation the same way: "Oh, Lawd, chile, guess what these niggas done did today?" And every phone call ended with: "I don't know how long I can stand it. 'Cause ya know I done raised mi' own churen. Chile, Lawd knows I don't need this mess . . . in mi' house. So I'm jus' gonna pray and believe dat de Lawd will bring me through. Ho, Glory! Thank ya, Jesus. Thank ya, Lawd!"

The part that really bothered me was her telling everybody that we were such bad young'uns, we'd tear up the Devil.

That weighed on me heavy as a gravestone. She and Reverend Pickett told us about the Devil, how he had a fight in Heaven with the God of everything and the Devil was still alive, but in Hell. Which meant we had to be the all-time worst kids on Earth, in Heaven, and in Hell ever. And for a long time I believed it, causing me considerable stress.

At school, the aftermath of the money-taking incident was that I was suddenly being teased by the other students—the same ones who'd sworn to me their endless friendship while sharing boxes of Good & Plenty, Lemon Heads, Boston Baked Beans, and Red Hots. Feeling scorned, without friends, no longer able to daydream, I lost all interest in school. Being there made me feel so uncomfortable, I decided not to go. This was easy to accomplish. In the mornings I'd come into school through the main door and go right out, leaving through another door. I spent the days playing by myself at the Tot Lot, a playground under construction, on the next street over from the Picketts.

Since no punishment loomed any worse than what I was already getting, I kept up the routine—slipping out from school in the morning and then, at lunchtime, returning to school for a little while before I'd leave again, going back just before the afternoon bell rang to go home. Or some days I'd go straight to the Tot Lot. Many weeks went by before anyone figured out that I was missing.

In the Tot Lot I watched the leaves turn colors, drooping from their branches till they gave in to the wind and fell to the ground, becoming piles for jumping onto or resting in. Under the constant gray threat of the sky, I felt the air change from moist and cool to wet and cold, bringing with it, finally, November's first snows. But the weather rarely stopped me from cutting school.

Sometimes I could get myself to think of little or nothing; other times I tossed and turned in my mind, mulling over everything that was happening at home in events, words, and images that came whirling around me like leaves being blown by the harsh winter wind.

I thought a lot about my fights with Dwight. It had been getting worse, too, the way Mizz Pickett treated Keith so much better than us, carrying on about him being so special because his father was white and Keith's skin was light and his hair was *"good."* That made Dwight turn his anger toward Keith, which, in turn, caused me to step in, fighting Dwight more than before, to keep Keith safe.

I stepped in because I knew what it was like to be the youngest and unable to defend myself. And I also fought Dwight because I wanted Keith to like me. In fact, I wanted to be like him. Later I could see that more clearly and see the irony that Dwight resented Keith for all the reasons I wanted to be Keith. Me, I was the ugly duckling, the outcast, and Keith was the opposite in every way, the golden boy. Even though he was younger, I thought that winning his approval would change my status somehow.

Dwight wasn't the only one I defended Keith from. There were others who picked on him, like the boy in the neighborhood who was pushing him around one day when I came out of the house, calling Keith a white boy and a honky. Keith's face was scrunched up in a sad frown till he saw me coming and shot back at the boy, "Here comes my brother, and he's gonna get you."

The boy glanced quick at me, then back at Keith, unconvinced by the difference in our appearances, and said, "That ain't your brother!" And he punched Keith in his face.

Keith gave a loud yell and burst out crying while I calmly walked up to the boy and punched him, *Pow!* in his face, knocking him to the ground, where he lay dazed.

Keith's tears stopped in their tracks and his face broke into a huge smile as he began laughing harder than I had ever seen him laugh before.

With encounters like these, I did succeed in winning Keith's approval. We became the best of friends. Even if he wasn't my real little brother, that's how it felt. Of course, this only made Dwight angrier, which escalated the fighting.

It was a poisonous cycle. We were on the same battle-field, each dodging what we feared most, retaliating in those same modes. With Dwight, everything was external and physical. His pain was written on his face and in the nervous jump of his movements. He was hands, fists, kicks, hitting, running, shouting, crying. With me, it was all on the inside. I was hands-in-pockets, watchful, quiet, and hidden tears.

Dwight hated being beaten, as if his body recalled in its cells every blow Mr. and Mizz Pickett ever dealt him. I hated the internal wounds, the words that said I was worthless and unwanted, the teasing, the ridicule that echoed in my ears and shattered my insides into dust. If given the choice, I'd have picked a beating over being shamed. But at least I found momentary relief in my in-ternal escapes. Dwight's only relief from his torment was to beat up on someone else. Since I was the closest to him, that was me.

With Dwight it was all about winning, and I under-stood why—because he could never win with the Picketts. His fights were his only survival skill. He had to win, he needed to win, and if it looked like he wasn't going to win, you'd be fighting him for hours. That was how he communicated his feelings, and so I fought him because he needed me to. Now and then I won, but only a handful of times.

So Dwight with his punches and me with my insides crushed to dust, we fought and fought and Mizz Pickett beat us and beat us for fighting, asking the question, "What's ill you niggas?! Why you fight so!" blind to the fact that she was the reason we fought. It wasn't just by pitting us against each other, there were times she physically encouraged it—even going so far as to come into the room where we were fighting and lean over with all her weight to hold Dwight down while I pounded him with my fists. I hated her doing that—holding him down so I could win and further humiliate Dwight, who would, for certain, make me pay later by fighting me with that much more fury. But since she was giving me my chance to beat him, I was afraid of what she'd do if I stopped. So I didn't.

Then one day she tried a new technique that we both hated, maybe even worse. Finding us in the midst of a brawl, she shouted for us to stop and took us each by a shoulder and pulled us apart. We stood there, still glaring at each other, panting.

Mizz Pickett said, "Now." Looked at me. Looked at Dwight. Said, "Yawl niggas, kiss."

Dwight and I swung our heads at her with big question marks in our eyes.

"You heard me. I wont-choo to kiss 'n' make up. Say ya sorry, too. Both o' ya."

The apologizing part was bad enough. But when it came to the kissing, Mizz Pickett wasn't about to accept halfhearted pecks on the cheeks. Instead, she folded her arms, pursed her lips, and commanded us to kiss on the mouth. When we did, neither of us hiding our disgust, the smirk I saw break out on her face was awful.

It was that grin of self-satisfaction that really got to me and made me keep playing hooky and keep taking the pennies and nickels I found around the house every so often. Then one day I made a tactical error. When I was playing at the Tot Lot, I saw a teacher drive by in her car and figured that school had let out early. Worried, I ran home, hoping to join up with the rest of the kids as they returned. My run slowed when I approached the house and didn't see any of them. Just as I was thinking about heading back toward school, I looked up and saw Mizz Pickett coming out of the house in her coat—probably to go back to the thrift store she owned and operated at that time.

I froze in a feeble attempt to become invisible.

"Nigga," she roared, "why ain't-choo in school?"

"T-t-teachers' conference," I offered, "they sent us home early."

Mizz Pickett looked like she was going to accept that. Then she squinted up her eyes suspiciously and marched me into the dining room where I had to stand by as she called the school. By the look on her face as she was talking, I guessed they were telling her that there wasn't a teacher's conference, and worse, about my absences. If there was any doubt in me after that, she whupped it out good.

The following Monday I went to school and was paddled by the vice principal. Then, when I arrived home that afternoon, I spotted Miss Blum's white VW Beetle out front and found her in the living room with Mizz Pickett. They looked up fast when I came in the room, as if they had been discussing me.

Miss Blum appeared to be upset, but in a kind, concerned way. Mizz Pickett's expression was just thrilled indignation. "Antwone," Mizz Pickett began, pronouncing my name like *Ant-too-ine*, almost with a French inflection, which was her imitation of what my social workers called me, "Miss Blum wonts to speak to you, alone." And she rose from the couch and walked a few paces away, but remained in the room.

Miss Blum told me she heard that I'd been skipping school and playing in the park. "Is it true?" she asked.

I nodded. From behind me, I could hear Mizz Pickett making a grunting, prodding sound. "Y-y-es, ma'am," I stammered, becoming more and more nervous.

Miss Blum asked, "Antwone, don't you get cold?"

It seemed a caring question, not what I expected. With effort, I tried to say that yes, I did get cold but it wasn't too bad. But the words coming out of my mouth were jumbled and inaudible. Miss Blum nodded without emotion and said that we'd talk more a little later. As I turned to go, Mizz Pickett sidled back over, sending Miss Blum a look of Well, what did I tell you?

On my way upstairs, I couldn't help overhearing Miss Blum say to Mizz Pickett, "He has a great deal of difficulty talking to me. It might be easier if I visited with him outside the home."

"You can get him to talk," I heard Mizz Pickett say, "but-choo can't believe a word he says." After a beat of silence, she went on, proclaiming, "He's the worst young'un that ever was, and the biggest liar that ever lived. And I always say—if ya lie, ya steal. We all know he's a thief." Knowing this well-used rant about me, I started on up the stairs but was stopped when I heard a new twist on it. Even though Mizz Pickett made it sound like a big secret, her voice was practically shouting as she said, "I heard through some family membas o' his that his daddy's a thief. It's in his blood, he got stealin' and lyin' runnin' through his veins, that word *thief* written right there in his blood."

Family members of mine? All of this time, Mizz Pickett had never mentioned that I had family members. Suddenly she was bringing them up. Family members that she knew who knew my father, a thief? I didn't know what to think or feel. Would Mizz Pickett make it up?

That's what it seemed that Miss Blum was wondering when she asked, "Surely, you don't believe Antwone actually has the word *thief* written in his blood?"

"Well, I don't suppose so," was Mizz Pickett's response to being molded.

The sound of papers being exchanged was heard, over which my caseworker explained to Mizz Pickett that I was to be taken to Metzenbaum Children's Center for psychological evaluation.

What that was, I didn't know. Not wanting to hear any more, I silently continued up the stairs and into the bedroom. A short while later, I heard the front door open and close. Afraid that she might see me peering out of my bedroom window, I stayed on my cot. But in my imagination, I saw Miss Blum bundled up as she hurried out to her white VW Bug, stopping at the last moment to look at the house with the backdrop of factories and smokestacks behind it, shaking her head sadly before getting into her car and driving off into the darkening day.

In the winter of 1967, I was seen by Mrs. Barbara Honhart, M.A., on two dates, December 13 and December 15, at the Metzenbaum Children's Center, an orphanage and child placement center. She made these observations:

Antwone is a small, dark complected
Negro boy who came to the first
session dressed very neatly in
white shirt and jacket. . . . He was
cooperative and responded fairly
well in the first session, but he
was quite sullen and stubborn during the second session. . . . It may
have been because after the first
session he had to wait two hours
before he was picked up by his foster mother.

Included in her report were remarks about an initial interview with Mizz Pickett, who said that I was "dis-
turbed" and "wild" when she got me, something that was
contradicted in earlier reports by social workers. After
detailing Mizz Pickett's accounts of my stealing and truancy, the report went on:

Mrs. Pickett elaborated that Antwone
never tells the truth about anything. For example, she has to check
him every morning to see if he has
underwear on because he will say he
has it on whether or not he does.
(Mrs. Pickett had discovered that he
was going to school without underwear on.) Another bad habit is that
he tears up and destroys things.
Antwone and Dwight do not get along
well. Antwone has "spells" when he
does not want Dwight to talk to him
or touch him. Mrs. Pickett related

that Antwone has never asked about
his name being different from her
and she has never explained to him
that he is a foster child.

Based on the various tests she gave me, Mrs. Honhart
described me as having the potential to function with
above-average intellectual abilities but that I was, obviously, not working up to that potential in school. Based
on what she termed "projective tests," she concluded that
I had a great deal of fear and anger:

Antwone told several stories about
murders and people drowning. . . .
His fearfulness may be a result of
his projection of his angry feelings, although there may be a reality basis to what he says about
Dwight and the boys at school beating up on him. Antwone seems to have
an image of himself as small and vul-
nerable and a wish to be indestruc-
tible. For the Draw-A-Person test he
drew a very large male figure whom
he called Dwight and a much smaller
figure standing on a stool who he
said was himself. . . . On the other
hand, he told a story about a man who
fell to the floor from a rope,
cracked the floor, but did not hurt
himself.

Some of Mrs. Honhart's interpretations were formed
without important information. She noted:

> Although Antwone expresses fear of
> being vulnerable and a wish to be in-
> destructible, he also expressed a
> desire to be a small child. He said
> on the Sentence Completion Test, "I
> want to be like my brother Keith."
> Perhaps he sees this brother as
> being protected by adults.

While it was true I saw Keith as protected by adults, that wasn't an indication of my wish to be a small child. The missing information was why I saw him as protected—because of his special status.

Mrs. Honhart did pick up on another problem area, what she said was castration anxiety and a fear of bodily damage. But without a way to ask me questions about Mizz Pickett, the beatings, or about Willenda, she could only guess at the causes:

> It is suspected that the circumci-
> sion this summer exacerbated his
> concerns. . . . Not only does he show
> concern about missing parts and dam-
> age to himself, he also had diffi-
> culty relating to the Rorschach
> inkblot with strong sexual connota-
> tions. . . .

Mrs. Honhart was, however, on the right track. At the conclusion of her report she commented:

> It appears that Antwone needs direct
> help in working through some of his
> feelings and conflicts, but it also

seems that the foster mother needs
some guidance around ways of han-
dling his behavior. . . . She seems
rather unresponsive to his fears and
concerns. . . . It also appears she
may be rather sexually stimulating
in her treatment of him. For example, she checks him each morning to
see that he has on underwear.

Between Barbara Honhart's notes and Miss Blum's, it
is clear that there was serious consideration about removing me from the Picketts' home. While they decided that
my removal from the only home I had known would be
more harmful than letting me stay, these outsiders were
apparently starting to question Mrs. Pickett's fitness to
take care of children.

Around this time, Miss Blum recorded another person's concerns:

Mrs. James, Principal at O. W. Holmes
School called me. . . . She worries
about the foster mother.

It's Christmas morning, ten days after I was seen by
Mrs. Honhart. It is still dark when we rise, me, Keith,
Dwight, and Flo. In breathless silence we rush downstairs, twelve-year-old Flo leading the way in her flannel
nightgown, Dwight and I following in our flannel pajamas, and Keith trailing behind in his one-piece sleeper
with feet that have plastic on the bottom.

Ever since summer, Mizz Pickett's been telling us that
we didn't deserve any presents at Christmastime and not
to expect any. We knew she meant it then but thought by

now she would have softened her position. And over the past week, with the snow piled high outdoors, Christmas decorations and lights all through the neighborhood, plus cooking and cleaning and holiday preparations for the past four days, we're caught up in the spirit of the season—and have forgotten her warnings. We have also forgotten, for the moment, that there have been many Christmases that have come and gone without a single present for us.

This year the Picketts have two Christmas trees. One tree, wrapped in multicolored lights, is out on the enclosed porch, where the wind has brought in a wet carpet of snow. The other tree is inside next to the television. Besides the lights, it is hung with different-colored bulbs and other Christmas ornaments. Underneath, wrapped presents have been accumulating, and this morning we spot several new presents we didn't see before.

The three of us boys stand back as Flo kneels down, looking for our names. But she finds gifts only for Keith, setting them down in front of him in a heap. We watch, Flo, Dwight, and me, long-faced and unhappy, as Keith, his handsome little-boy face beaming, his wide brown eyes bright, his goldish/brownish curls bouncing, eagerly opens them—trains, cars, balls, horses, cowboy and Indian toys. Keith starts to play and then looks up at me, asking, "Twonny, you wanna play with some of my toys?"

"No, that's all right, Keith," I reply, not wanting to make Dwight, Flo, or myself feel worse about not getting anything. "I'm gonna just watch TV."

Keith shrugs, saying, "Okay," and goes back to playing as the three of us sit down and watch a Christmas cartoon about Mr. Jingaling, the keeper of Santa's toys. Actually, we're mostly watching Keith and his toys.

The next person up is Mizz Pickett, who comes padding into the room in her robe and slippers. She looks first at Keith with an approving eye, then spots us and asks, "What yawl doin' down here so early for? Ain't nothin' under that tree fuh ya." Her eyes now flash that familiar smirk that says—*I told ya.*

Over the remainder of the day, we don't complain to each other. After all, we should be used to this by now. By mid-afternoon, the house starts filling up with Mr. and Mizz Pickett's various kids and grandkids, plus some church friends.

All day more presents have arrived and been exchanged. Each arrival has ushered in more food to be laid on the table, which by four o'clock is blanketed with cakes and pies and roasts and casseroles. My attention goes right to the sweets—pound cake, chocolate layer cake, coconut cake so covered with shaved coconut that my secretly taking a pinch of coconut goes undetected. There are pies made of pumpkin, cherry, blackberry, rhubarb, and sweet potato. There are bowls of chestnuts and striped Christmas candy and candy canes.

Aside from not getting any presents, we aren't excluded from the festivities. And yet, it's clear to me that this holiday, like all their family gatherings, is for the Picketts, not us, as if we just get to eat because we happen to be here and it would be rude not to offer us something.

Although little will change this observation over the years to come, this Christmas does bring with it a surprise. In the night, a half hour after Mercy has arrived, she motions me to her in the living room. She is smiling, with the warmth and kindness in her eyes that I have come to know in her. And then she extends her hand, holding in it a gift for me, a flat package wrapped in cheery Christmas paper.

I don't know what makes me happier, the feeling of tearing open a present that Mercy thought enough of me to pick out and buy, the only one out of us she gave something to, or the gift itself—a tomato red shirt with a mustard-colored stripe across the chest, across the shoulders, and down the sleeve. My words of thanks are soft, but I am sure Mercy must see my joy, hopefully enough to hold in her heart, to know that her thoughtfulness made one of the only real Christmas days of my childhood a memorable one.

T he only thing that separates me from my father is a wooden door. That's what I'm thinking. I'm in the waiting room just outside the office where he is. In ten minutes, the door is going to open and there he'll be, my father, Dr. Fisher.

A wall clock ticks above the reception window in this small alcove of a waiting room. I'm sitting with Miss Blum and Dwight, he and I dressed in our good clothes that are too small for our growing bodies. It's the end of March 1968, no longer winter, not yet spring, raining hard outside.

Dizzy with too many feelings to sort out, I try to understand how this momentous occasion has come to be. Once we're together, my dad and me, I'm sure everything will be explained and solved. Nothing can be perfect, I know, but just to meet him finally I'll be too happy to care.

Even though I've stopped playing hooky at the Tot Lot, my problems at school and the fights with Dwight haven't changed much. Christmas made Dwight miserable—Keith getting lots, me getting the shirt from Mercy, and him nothing. Lately, I've been spending even more time with Keith, and that's made Dwight all the more angry.

Keith had recently started to go visit his real mother on the weekends. Up until this point, he hadn't been told about her and probably assumed that Mizz Pickett was his mother, just as we once did. When he came home looking confused and upset from his first visit with his real mother at social services, I asked him how it was, but he didn't want to talk about it.

On Sunday morning of that same week, Keith, Dwight, and I were in our room, dressed and ready for church, when Mizz Pickett called from downstairs, "Twonny, you and Keith come on down here."

Telling Keith, "Come on," I led the way. Mizz Pickett was waiting at the bottom of the steps, dressed in one of her fancy Sunday-church-meeting outfits. Reaching down, she adjusted Keith's clothes and then took us each by a shoulder and guided us into the living room. Ahead of us was a woman who looked something like a slightly heavier version of a young Aretha Franklin. Seeing her, Keith tried to turn and run, but Mizz Pickett kept a firm grip on him, saying, "Your mother wants to take you out, Keith."

"I don't want to go!" Keith insisted.

Fanta, Keith's mother, reached out her hand to him, saying nothing.

I didn't understand what all this had to do with me. Then, like a prayer being answered, Mizz Pickett added, "Twonny's goin' with you."

Fanta smiled knowingly, as if this was part of a plan. What popped into my mind was: Wow! That means I ain't goin' to church! Anything to get out of going to Mr. Pickett's House of God, I thought fast, and said brightly, "Come on, Keith. We're gonna have some fun!"

Mizz Pickett gave me a look, as if to say, Look a here,

nigga. Tone it down, ya actin' wit' jus' a little bit too much zeal.

As out of reflex, I put my head down, deciding the situation didn't need any help from me.

Fanta offered, "Let's go, boys. We're gonna go shopping, downtown."

With this, Keith's face went from worry to hope as he turned to me, saying, "You gonna go with me, for real?"

With trepidation, I looked at Mizz Pickett, and she answered for me: "Yes."

Keith grabbed me by the sleeve, pulling me toward him.

It seemed like an hour to walk from where Mizz Pickett was standing to Fanta's car in the driveway. Once inside it, I felt all that feeling of an inmate on leave. Fanta treated me nice, buying us those big salted pretzels and sodas, and some clothes for Keith. Just like I promised him, we had fun and he warmed up to his mom as the day went on.

After that visit, Keith never wanted to go anywhere without me, and I was satisfied with that. I was happy Keith was getting to know his mom. He was such a good-looking, sweet-natured boy, I couldn't understand why she wouldn't want him to live with her all the time.

My going out with Keith on visits wasn't Mizz Pickett's idea. Apparently, Keith had said something to his social worker that had caused her to make the suggestion. Then, through the grapevine of Mizz Pickett, from my social worker, came an earth-shattering suggestion for me. Because of my bad performance in school and the constant fighting with Dwight, it was decided that we were going to be taken to see a man by the name of Dr. Fisher. That's all she said, giving me two days to think of nothing but him.

Fisher. Same last name as me. I kept rolling that name in my head, *Dr.* Fisher. For the longest time, my last name had mystified me. Mizz Pickett did her grocery shopping at Fisher's Supermarket and whenever she took me with her, I'd scan the aisles and the checkout area looking for possible family members. Standing in line, it used to embarrass me when it was time to pay and she'd reach into her blouse for the money she kept in her bosoms—that's what she called 'em—and, while her hand was there, rearrange herself, too. Just in case any of my relatives named Fisher happened to be there, I was sure they'd want nothing to do with her or me.

Then I was told that I was going to see Dr. Fisher. He was the only person I ever heard of with my last name. It had to be *him.* Mizz Pickett had said he was a thief, which she knew from family members. A thief *and* a doctor? It didn't make sense. But I guessed the day had come when my being the worst young'un that ever lived had become so impossible that they were sending me to my father so he could give me a good talking to.

So here I am, in a chair sitting outside of his office, determined to be as nice as I can, thinking maybe he'll like me and think about keeping me. Inside, to myself, I make a promise, like a prayer, that if my father does keep me and I don't have to go back to the Picketts, I'll be good for the rest of my life.

The instant I make my silent pledge, the receptionist comes around and motions to Miss Blum, who escorts me and Dwight to the door. Miss Blum opens the door, gesturing for us to enter, and then closes it behind us. We're alone in his office. A man in his late thirties or so, old enough to be my father, but not too old like Mr. Pickett, he's sitting at his desk, wearing a gray suit, white shirt, and tie. He looks at us, smiling, and says, "Hello, I'm Dr. Fisher."

Dwight asks if he can look at the toys in the toy box and, when he gets permission, hurries over to them. I remain silent, hummed into this man, my eyes locked in on his presence as he stands and walks over to me. With his hand on my shoulder, he guides me to a seat in front of his desk. I hop up into that chair, eager for what comes next. Other unsure feelings, though, are starting to flutter through me, and I'm not ready for them. So I try to go blank, like I know how to do, and turn my attention to the room itself.

The desk sits at the side of the window. To the right is a couch, adjacent to the toy box where Dwight is playing. I'm about to turn my gaze the other way when I focus in on a painting on the wall. It caught my eye before when I hopped on the chair, just as a splatter of colors. Now I notice that it's a picture of something. Studying it, I can feel him studying me. That's okay, I think. I'll look at this painting a little while longer and decide whether this man is really my father.

Then he speaks. "What do you see?" he asks. Not the question I'm expecting.

"A rainy day."

"Why do you think it's rainy?"

" 'Cause it looks . . . wet."

"Is that all you see in the picture?"

I look again and see a story in the rain. It's a street with buildings on both sides, streetlamps, and cars driving up and down the street. I tell him.

He nods, pleased. Again he asks, "Is that all you see?"

Then I think, no, there's more. I see you don't look like me and I don't look like you, either. You look more like my social worker—she's white, too.

Even though I came to the conclusion that Dr. Fisher wasn't my father, why we had the same name remained a mystery to me. In any event, over the next few months that me and Dwight were sent to him, the sameness of our names made me come to think that this man was some sort of an ally.

According to Miss Blum's notes, the plan was for me to receive a steady four to six months of therapy, then she would ask Dr. Fisher for an evaluation as to whether I should stay in the Pickett home or be moved. Early on, Dr. Fisher informed social services that Mizz Pickett should receive therapy as well. Before transferring my case to a new social worker, Miss Blum made reference to both our visits and Mizz Pickett's, noting:

> Mrs. Pickett is somewhat lax in keep-
> ing appointments and caseworker must
> constantly remind her of their ex-
> treme importance.

I believe that Mizz Pickett never intended to maintain the visits and probably was angry when Dr. Fisher took my side about the issue of stealing—after Dwight and I both explained our occasional "theft" of ten or twenty cents as the result of never being given any money. But not wanting to appear to be uncooperative, Mizz Pickett called Miss Blum to say:

> She has made Antwone a bank at the
> suggestion of Dr. Fisher and has
> given it to him. Now she gives him
> work around the house for which she
> pays him so that this bank is his and
> he takes care of the money and also

> does not take money from other peo-
> ple. Mrs. Pickett seemed very proud
> that this has worked well and has
> done a lot for the child.

I remember it somewhat differently. She did make me a bank from a tin can of Clabber Girl baking powder and put it up on her dresser for safekeeping. I decided one morning, that since the money was mine, I would use some of it to buy myself some candy on the way to school. But when Mizz Pickett discovered what I'd done, she let me know vividly that I wasn't allowed to take that money. That night on the phone, she told everyone what I'd done this time, exclaiming to each person, "Lawd have mercy, this nigga's stealin' from hisse'f!"

Some years later, Dwight and I found out some surprising news. It came out one day after Dwight said something directly to Mizz Pickett about how hard it was never having any money, and she shrugged it off, saying, "They don't give me but five dolla's fo' ya allowance."

Dwight and I couldn't believe it. If we were supposed to be getting an allowance, we never got to touch a dime of it ourselves. Not long after she made that comment, Mizz Pickett had to go down south on a visit and had Mercy come stay at the house to look after us. At one point, Mercy had left some change out on the table. It was all of a dollar and I took it, hurriedly leaving the house to go somewhere and spend it. When I came back, Mercy was in the kitchen cooking. She greeted me warmly, as she always did. I was sheepish, already feeling ashamed for taking money that was hers. Then, as if she just remembered, Mercy said, "Oh, Twonny, I have your money. I put it there on the table." She nodded in the direction of where I took her money, indicating that it was mine from social services.

"My money?"

"Your five dollars. Go on in and take it."

"I don't need no money," I mumbled, trying to hide my shame.

"Go on and take it," Mercy insisted, turning her attention back to the delicious-smelling dinner she was making.

There was no refusing. She made sure I went in and took the five dollars, more money than I'd ever received in my life. It killed me to take it, it killed me a thousand times. Mercy didn't smile. She didn't look at me knowingly. But, of course, she knew everything and was wise enough to give me the real lesson her mother never had, the only lesson I ever needed to learn. And that was the last time I ever stole anything in all my life.

I believe Mizz Pickett only told Miss Blum about the bank and that I was making progress because she felt that both the social worker and Dr. Fisher were getting too close to the truth, and she wanted them to stop prying into the situation at home.

My next caseworker, Miss Haller, went on to write that although Dr. Fisher had attempted to offer Mizz Pickett what they called "back-door therapy," she had failed to keep those appointments in addition to mine and Dwight's. Now Mizz Pickett changed her approach. The report shows that her excuse, when questioned, was to claim that therapy wasn't helping and that I was getting worse. In fact, she asked for me to be removed from the home. But as soon as Miss Haller began taking the necessary steps—and asking the necessary questions—Mizz Pickett reversed her position:

    It wasn't long after she asked this
    that she said they wanted him to stay.

She said she and her husband could
never think of letting him leave.
Antwone appears to accept the foster
parents and his foster sister and
brothers. Antwone has become much
more verbal in the past year and
seems to be able to stand up for himself. He is involved in activities,
having joined the Boy Scouts.

While Mizz Pickett had momentarily convinced social
services that the visits to Dr. Fisher weren't needed, Miss
Haller later attempted to reinstate them, but to no avail.

In the meantime, I was on my way to failing the fourth
grade. My report card described me as, "normal in class,
a loner with children, and dependent." I received a C in
social studies, Ds in reading, spelling, math, handwriting,
and English. In phys. ed., music, and art, I received Bs.
But it was only art that gave me the feeling of being able
to do something reasonably well that I genuinely enjoyed. This was around the time that the other kids started
asking me to show them how to draw things—like Santa
Claus. Their Santa Clauses were more one-dimensional;
mine had a big smile that disappeared into his beard,
grinning eyes, rosy fat cheeks, and a hat that slung over
sideways. Their houses were flat; mine were at an angle
and multidimensional. Their suns and skies were only at
the top of the paper; my sun and my sky reached down to
the ground, in perspective. No one had ever taught me to
do this. I was just drawing it the way I saw it.

Even Mizz Pickett was complimentary of my talents,
going so far as to encourage me to enter one of those
magazine contests that asked you to finish drawing
Bambi. When the response came back that I showed def-

inite artistic talent and was eligible for their art correspondence course, Mizz Pickett was personally delighted. Of course, it cost money that no one was offering to pay. Even so, for those twenty minutes on that one afternoon, Mizz Pickett did not claim that I was the worst young'un that ever lived. She actually bragged about me to a friend who asked me to draw Jesus.

Fleeting though it was, it was my first taste of doing something that gave other people pleasure, and I liked it.

There was a change going on already in my awareness about others and their concerns. Until now, my only sense that there was a bigger world outside my own existence came through television.

First it was the war. Willenda watched it every evening. When she was around, it was a religious practice at the same time every day for her to enter the living room and change the channel from whatever we were watching to the evening news program of Chet Huntley and David Brinkley. That was the show that brought us the war in Vietnam.

Although many families in the Glenville area had been hit hard by the war—losing sons, brothers, fathers, and boyfriends, or having them come home handicapped, drug-addicted, shell-shocked, and traumatized—I wasn't aware of it yet. The Vietnam War, for a boy like me in those days, seemed glamorous. From what I could tell, war was a job. Sponsors of Saturday morning cartoons advertised war toys. There were lifelike M-16s, M-14s with bayonets, pistols that shot plastic bullets and popped like firecrackers, bazookas, grenades, and camouflage army fatigues. And I watched the television shows *Rat Patrol* and *Combat* with eager delight, never realizing the awfulness of war, the dreadfulness I'd come to understand in later years.

The news program showed the other war as well, the

social war here in America. This was the one that made me worry. People who looked like me filled the Picketts' twenty-inch television screen, all of them marching in huge crowds and appearing at odds with police, who were sometimes shown beating them. Firemen even sprayed them with water hoses. All I knew to think was what I had been told—that the policemen and the firemen were my friends and there to help me if I ever needed them. That meant all those people in the crowds had to be trouble-makers. But the main guys causing most of the trouble were two black men who seemed to be the leaders of the crowds. Their names were Ralph Abernathy and Martin Luther King. I wished they would stop making trouble and was convinced if they would go away, the police wouldn't get so mad at the people who looked like me and everyone else in my own smaller world.

Ironically, I knew nothing about the racism that was at the heart of this war. That came from being insulated—living in an all-black neighborhood where nobody was calling me "nigga" but Mizz Pickett. Cleveland at this time, in fact, had become the first major urban center to boast a black mayor, Carl Stokes. I didn't know that Martin King had come to Cleveland and endorsed Stokes. All I knew was that when they finished building the Tot Lot, Mayor Stokes came to our neighborhood for the dedication and gave away basketballs.

Neither the Picketts nor anyone else ever explained what King and Abernathy were trying to do. No one explained that these were good men. I saw them many times on television being escorted in and out of jail, and because to my knowledge only bad people went to jail, that's what I thought they were.

And then came the night of April 4, 1968. According to my social services record, Mizz Pickett had taken me and Dwight that afternoon to Dr. Fisher, one of the last appointments we made. She was in the kitchen talking and laughing with a handful of other adult friends and family members. The evening news was over; Willenda had abandoned her TV-watching chair and left me on the floor in front of the set to watch by myself.

All of a sudden the screen went blank. A voice came on saying that the program had been interrupted for a special announcement. When the next voice announced that Martin Luther King had been shot, I jumped to my feet and headed for the kitchen, proud to carry the happy news to Mizz Pickett. The instant I spoke, confusion erupted as everyone pushed past me to the living room to hear it for themselves. From where I stood in the kitchen, I heard Mizz Pickett let out a howl unlike any I'd ever heard, even to this day. It was terrible. When I walked back into the living room, everyone was crying. As the bearer of the news that made everyone so unhappy, I felt responsible.

Things got worse. The city literally exploded into fire. With widespread rioting, we were sent home early from school the next day. On my way home, I saw army Jeeps and trucks speeding up and down 105. My pace quickened with every step, certain that the world I was now walking through had become like the one in Vietnam and on *Rat Patrol* and *Combat*. Before I knew it, I was racing at top speed back to the Picketts' house, the place where I felt all of this had started. I flew upstairs to the bedroom. There, I slid underneath my cot and waited for the bombs of war to fall.

A few months later, I awoke to the morning after Robert Kennedy was killed. The Pickett house was filled with sadness. Everyone kept mentioning Martin Luther King and Robert Kennedy in the same sentence, charging the air with an unbearable weight that pressed down on my shoulders and around my heart. It was obvious that I was responsible for Robert Kennedy's death, too.

Until that day, I only understood my own sadness. But all at once, I became aware that there was more sadness out there than I could even begin to imagine, enough to fill every ocean full of tears. With too many tears of my own to hide, I ran outside, out to the backyard, behind the garage. Then I leaned my head in my hands and cried for a long time.

Same summer, not much time later, close to my ninth birthday, it's a muggy, overcast Saturday after cartoons, after housework. The smell in the air promises rain, but judging from the medium gray of the sky, I think I can make it to Mr. Murphy's and back before it starts to fall. No one around upstairs, I steal another nickel from myself out of my Clabber Girl baking powder can bank in Mizz Pickett's room and slip out of the house unnoticed. Barefoot, dressed in a crewneck shirt and cargo shorts with zippers on the pockets, I run up our block and over to 105. The thunder rumblings begin. Just God moving around his furniture, that's what Flo likes to say, even though the big booming sound scares me a little.

At Mr. Murphy's I buy five Smarties rolls, little sweet tarts wrapped in cellophane, and unroll them into my pockets, leaving one pocket unzipped enough to stick two fingers in so I can eat the candy piece by piece on the return trip.

Halfway home, the sky goes from dark gray to almost black and a loud thunder snap accompanies the first few raindrops that fall. Heavy, warm, big drops, they drench me in seconds, like an overturned bucket from the sky dumping just on my head. I reach my hands up and out, as if that can stop my getting wetter, and open my mouth, trying to swallow the downpour, till it finally hits me how funny it is, my trying to stop the rain.

This is so funny to me, I laugh and laugh, as loud and free as I want. Instead of hurrying to higher ground, I jump lower, down off the curb, splashing through the puddles, playing and laughing all the way home. In all my life till now, rain has meant staying inside and not being able to go out to play. But now for the first time I realize that rain doesn't have to be bad. And what's more, I understand, sadness doesn't have to be bad, either. Come to think of it, I figure you need sadness, just as you need the rain.

Thoughts and ideas pour through my awareness. It feels to me that happiness is almost scary, like how I imagine being drunk might feel—real silly and not caring what anybody else says. Plus, that happy feeling always leaves so fast, and you know it's going to go before it even does. Sadness lasts longer, making it more familiar, and more comfortable. But maybe, I wonder, there's a way to find some happiness in the sadness. After all, it's like the rain, something you can't avoid. And so, it seems to me, if you're caught in it, you might as well try to make the best of it.

Getting caught in the warm, wet deluge that particular day in that terrible summer full of wars and fires that made no sense was a wonderful thing to have happen. It taught me to understand rain, not to dread it. There were going to be days, I knew, when it would pour without warning, days when I'd find myself without an umbrella. But my understanding would act as my all-purpose slicker and rubber boots. It was preparing me for stormy weather, arming me with the knowledge that no matter how hard it seemed, it couldn't rain forever. At some point, I knew, it would come to an end.

three

*C*ome *to Jesus, come to Jesus, come to Jesus, right now"* were the words sung to a plaintive tune toward the end of every church service. Standing there in the pulpit, sweat dripping off his shiny bald head, Reverend Pickett led the singing, his thick, stocky arms outstretched from his thick, stocky body as he beckoned us to come to him. I couldn't understand why the image of Huckleberry Hound came into my mind every time we sang this song—that is, until I realized that we were singing to the tune of "Clementine," Huckleberry's song. The image of a cartoon dog didn't seem to go with the visual of Reverend Pickett standing there in what I assumed was the way Jesus might gather his flock close.

At the Holy Temple Church of God and Christ, Mr. Pickett's storefront church on Woodland Avenue, his flock was considerably smaller in size than in earlier years. There was Mizz Pickett; the blind man who played the piano; the ninety-year-old mother of the church, whom we called Mother Crump; plus us kids. Every now and then, there were one or two others who might have been corralled into attending that day. Otherwise, it was just the six or seven of us. We were the congregation, the choir, and the musical accompaniment.

This never seemed to bother Reverend Pickett. He'd look around at the sixty-odd empty seats, then at us, and say, "The Lord said wherever there are two or more gathered in His name, there He will be also." He punctuated that with a quick raspy clearing of his throat, followed by, "Praise God," which was our cue to say, "Hallelujah," "Thank you, Jesus," or "Yes, suh."

Every Sunday, when we drove to the church on Woodland Avenue, Mr. Pickett was expectant and ready—as if this was the day that all Cleveland's sinners would turn out to be saved, filling the storefront to overflowing. That must have been his vision when he bought this particular building in its more run-down neighborhood outside of the Glenville area. There were apartments upstairs that were rented out and two storefront spaces downstairs, one of which was sublet to various tenants and the other where he installed the Holy Temple Church of God and Christ. Mr. Pickett built everything himself—the pulpit, the podium, the pews, even a choir loft. For the two front windows, he covered the inside of the panes with Con-Tact paper made to look like stained glass and, with a razor blade, cut out the middle in the shape of a cross, and replaced that Con-Tact paper with white construction paper. If you saw it from up close, you could tell all of that. But from a distance, it sort of looked like a real stained-glass window with a glowing white cross at its center.

Mr. Pickett bought the building around the same time, in the summer of 1969, that we were informed, suddenly, that the Picketts were moving to a different house. It was a two-story Victorian house in a better neighborhood of Glenville, two houses from the corner of Parkwood Avenue—enough of a distance from the old house that I had to change elementary schools from Oliver Holmes to Parkwood Elementary. Given only two days' warning before the move, I was filled with panic at the prospect of having to deal with a different neighborhood, different kids, and a different school, on top of the fact that I was going to be repeating the fourth grade. My fears were shared by Keith, Dwight, and Flo. It didn't take long, however, to adapt to our new surroundings and to make new friends in what was a nicer neighborhood than the one we'd come from. The houses were bigger, the kinds with high slanted roofs for all the rains to run off, and the lawns wider and greener.

Best of all, our block was lined by an assortment of trees with branches that met over the center of the street. Though there weren't many flowers around, in spring the leafy canopy seemed to make up for the lack of color, and I enjoyed it. I appreciated the shade it gave me in the summer and the way the icy bare branches looked like they were holding hands in the winter.

Ulysses Pickett, Sr., and his wife, Isabella—named Tha' Lady Isabella for a regal touch—appeared to be moving up in the world. In the house on Drexel Avenue, Mizz Pickett was so proud of her new formal living room that from day one we children were not allowed to enter through the front door of the house, but had to come around to the side and enter through the kitchen. Mr. Pickett must have been feeling successful, because a short while later he went out and bought himself a used white Cadillac. An earlier model, it had big silver fins and was all electric, like a rocket ship.

For the drive to church on Sundays, we continued to ride in Mr. Pickett's old blue Plymouth station wagon, stopping twice on the way—first to pick up Mother Crump and then the blind piano player. The drive should have taken only twenty minutes at the most, but it was often an hour by the time we got to church.

The service was almost always the same. Singing came first, then the offering, then more singing. After that came testimony from the membership, then Mr. Pickett would announce it was time for a song from the choir. As Keith stayed with us at the new house only for a few months before going to live with his mother, that made the choir just the three of us—Flo, Dwight, and me. We would get up from our seats and file past Reverend Pickett and traipse up to the choir loft in the back and sing. I could see Mother Crump in the first row, part deaf and trying with difficulty to follow along, the blind piano player off to the side, Mizz Pickett just behind him, and the back of Mr. Pickett's head as he rocked it side to side in rhythm to our singing. After we finished to the applause of three, we filed back down to our place in the front row by Mizz Pickett.

All of this was to build to the preaching by Reverend Pickett, which would culminate with the singing of "Come to Jesus," where each of us were expected to go to him, one by one, to be prayed for and to ask for forgiveness for our sins. Then there was another opportunity to make an offering, a final song, and an end prayer asking the Lord to keep us safe till we met again. And, finally, before we trucked home, we were given cookies.

Dressed in her finery, Mizz Pickett sat next to us, one eye proudly on her husband, the preacher, the other eye cocked sideways to watch our every move. One of her hands was poised to reach out and poke us when we did something wrong and her other held a large woolly-headed mallet. This she used to beat the big bass drum at her side. It impressed me what a limber wrist she had, how she kept the beat on the drum so well without looking.

When it came to her testimony, Mizz Pickett testified the same every Sunday, standing up to face the preacher, from time to time turning around to acknowledge us. She'd begin, "I jus' wanna thank the Lawd fo' bein' here and fo' wakin' me in my right mind, with the blood runnin' warm in my veins." This was said beautifully, like a happy song. But then her face would cloud over, the corners of her mouth turning down as she said, "I wont-choo to pray for my wick-ed chu'ren." Her tone when she said the word *wicked* made me think it was the worst cuss word there was and she was calling it of her own children. "Pray for them," she'd continue, " 'cuz they was all raised in the church. And now they out shackin' and . . ." Here she'd slow, as if to select what worse sins they were committing, but simply summed them up as "and what-not and somethin' other." At this point, Mizz Pickett would make a sweeping gesture with her right arm toward us, a serious frown on her face as she said, "And pray for these here chu'ren." Then brightly, she'd add, " 'cuz ya know, a *child* saved is a *soul* saved," pausing for a beat as she looked right at you, eyes flashing, before she threw in the bonus, adding, "plus, a life."

The actual author of the "child saved" line was a preacher from Alabama named Elder Taylor. A slick, high-energy type of preacher, he wore a black bowler and meticulously pressed clothes, and acted like he was some kind of a gunslinger for the Lord, like a baptismal sheriff. He badgered everybody who wasn't baptized, including me, into letting him do it. No one escaped his persuasive abilities.

When he preached, Elder Taylor played an electric guitar to add to the drama of his presentation. To me it was almost funny because he knew only one chord. But to Reverend Pickett, it was such a good idea that he went out and bought himself an electric guitar—as if hoping to liven up his preaching. But, fortunately, since he only learned to play one song, "The Star-Spangled Banner," Mr. Pickett never went so far as to play it in church.

Each of us kids had to rise for testimony and we said the same thing: "Thank you, Lord, for my mother, father, sister, brother. Please pray for me," and we delivered it with the same unemotional, fast delivery. Of course, there was never a question about us not doing what we were expected to do in church. In fact, the mere idea of questioning anything at church was blasphemous. And that was the terrible part because since I had questions about everything, I felt sure that I was truly bad. Still, the more I paid attention to Reverend Pickett's sermons, the more confused and scared I became.

The thing that bothered me the most was probably the resurrection. Mr. Pickett's eyes would light up and his body would tremble when he talked about the joyous day of the resurrection—Judgment Day—when the waiting and the suffering would be over. Maybe it was because I was watching *Dark Shadows* on television every day, but in my mind the resurrection would turn the streets into huge cemeteries with ghosts busting out of their graves and shooting up to Heaven or being sent to eternal Hell.

I didn't want to be judged, not the way they told us about Heaven cracking open and God and Jesus lifting you up to carry you home. Besides that, I never understood how the poor black folks in Cleveland like us ended up in a world with a white God and his white son over it all. And what if Judgment Day came at night? I could just see it, how there would be this huge explosion and the sky would split open. And God would roll out in this big white marble chair—kind of like the one Lincoln sits in at his memorial in Washington—and Jesus, the way the Picketts described him, would be on the right-hand side of God. I saw that Jesus would be kind of leaning on the chair because I imagined that he would be cooler and hipper than God, being younger.

After the chair rolled out, I worried that out of all the people in the world, God's spotlight would be on me, first. Why me? Because, like Mizz Pickett said, I was the worst child that ever lived. The bad part wasn't being judged; but to be judged first with everybody looking at me, that would be the worst.

In church, life was talked about as if it were just a waiting room and not to be enjoyed. The joy would come in death—but only if you were saved. Dwight got saved. That was the time when he was eleven and I was ten, in the summer of 1970, that we tarried on the Lord and Dwight caught the Holy Ghost, making it a two-for-one.

For a week leading up to that Friday night, Mr. and Mizz Pickett had been preparing us, telling us in just so many words that because we were so rotten and regular church wasn't curing us, we were going to go stay up all night and tarry on the Lord.

After dinner, Reverend Pickett changed out of his khakis and into one of his Sunday suits and loaded up the station wagon, driving us, along with Mizz Pickett and Elaine, one of the older daughters, to the church. The building tended to have a damp smell of mildew that was stronger after it rained, as it did that muggy summer night. Lit by the hanging bulb with its narrow lamp shade, the church looked more slapped together than usual, like the places in the ceiling where Mr. Pickett hadn't finished putting up the acoustical tiles. It wasn't the most comfortable setting in which to wait to catch the Holy Ghost.

Elaine left us to go wait back in the entrance area, behind the swinging saloon doors that Mr. Pickett had installed in what I thought was his attempt to make coming to church a festive affair.

114

Moving like a drill sergeant, Reverend Pickett set up three metal folding chairs in front of the altar and instructed us to kneel in front of them with our elbows on them, hands pressed together in prayer position, and to repeat the word *Hallelujah* over and over. We were to repeat it slowly at first, then faster and faster.

When we began, Mizz Pickett took a seat right behind us, in the front row, coaching and cheering us on, saying, "That's right, Twonny, you go 'head, that's right chile, go 'head on. That's right, Flo, tarry on 'im, you gone catch the Holy Ghost! De Lawd's gonna save you! Let Him bring ya on inta His fold!"

She kept it up for about two hours, adding in whoops and hoots, hollering to all of us, "Come on, chu'ren, yawl gone get the Holy Ghost and ev'rything gone be all right with God!"

In the middle of her biggest whoop, I popped up from my knees and announced, "I g-g-got to go to the bathroom." Before she could say anything, I turned and started on my way, feeling her eyes like red coals burning my behind as I walked up past the pulpit, by the piano, and into the little bathroom. Once inside, I rubbed my knees, stretched, and looked around, having lost the impulse to pee. Lo and behold, someone had left a *Jet* magazine in there, probably someone as bored as me, or at least more resourceful. I started flipping through the pages. The big news was about Idi Amin and all the terrible things he was doing in Uganda. So I killed as much time with Idi as I could, then played Green Hornet and Kato jumping off the toilet seat before heading back to resume my tarrying.

Moments after kneeling down and starting up my hallelujahs, I heard a sudden rumbling, like a pile of chairs toppling over. Judging from Mizz Pickett's euphoric shouts at Dwight, I understood that he had caught the Holy Ghost. Maybe I felt I'd get in trouble if I looked, or maybe I thought it should be a private thing, but I kept my head down and continued chanting. Meanwhile, Mizz Pickett was so excited, crying for joy, like a child was being born right in front of her. And, for the Picketts, it probably was like that. Or, as they would say, being born again. They were so thrilled with Dwight's catching the Ghost that they decided we would pack up and go home. Kind of like they'd been out fishing and were satisfied even if they got only one fish.

When we walked out to the car, I stole a glance at Dwight. Even though I suspected that he had just gotten tired of tarrying and decided to imitate the way he'd seen other people at church catch the Ghost, he really looked like something had happened to him. Like somebody just whupped him with a rope. He also seemed to be embarrassed, because he wouldn't look at me.

Dwight became even more sheepish in the car as the praise started heaping on him from Mr. and Mizz Pickett and Elaine. "Boy," said Reverend Pickett to him, "you know you goin' to Heaven now."

Mizz Pickett joined in, "That's right. We gone see each other in Heaven and walk us some streets of gold, you hear me, Boy? We just gonna have a time in the Lawd."

Elaine reminded Dwight how Jesus had said in His father's house there were many mansions and He was going to prepare a place for the people who were saved. As she was saying this, Mizz Pickett jumped back in, "Dwight, you gotcho' mansion now. It's bought and paid fo', you just got to stay out of the devilment. No backsliding, now, cuz the Lawd has a place fuh you!"

Flo was going along praising her brother, too, and they were praising her for being proud of him. Me, I didn't say a word. It was implied that I was on a fast track to Hell. Just in case I'd missed that point, Mizz Pickett nudged Elaine and nodded disgustedly at me, saying, "I ain't never seen nobody tarryin' on the Lawd get up and go to de baf'room, come back and tarry some mo'." She shook her head in sad amazement, repeating, "I ain't never seen that. I never seen nuthin' like it."

When the praising part settled down and they started talking about other things, Dwight finally turned to look at me.

"What happened?" I whispered.

Dwight shrugged and got a spooky look. "I don't know," he admitted.

Over the next weeks and months, I waited to see if getting saved made any noticeable changes in Dwight. To my eyes, he was the same, if not more agitated and fighting more of the time. This did nothing to change my confused, fearful feelings about going to church. So I began to tune out as much of it as I could. When it was time for a selection from the choir, I would stand and appear to sing along with Dwight and Flo, but only my lips would be moving.

Then, one autumn Sunday, when we start back from the choir loft, Dwight hisses in my ear, "You ain't singing, Twonny. I'm gonna tell."

Mizz Pickett narrows her eyes, shutting us up, and we walk silently past her to our seats. During the sermon and the climactic "Come to Jesus," my mind roams and I don't budge from my chair, even when Mizz Pickett, Mother Crump, Flo, and Dwight get up to be blessed by Reverend Pickett and anointed by holy olive oil from a bishop in Ypsilanti, Michigan. During the second offering, I occupy myself counting the three and half dollars that's in the plate, including the change that Mizz Pickett gave us kids to put in. Considering that and the fact that Reverend Pickett pays the blind piano player three or four dollars to play, I can see that the only paying customer is Mother Crump.

When it comes time for the final song, I'm so lost in thought that I don't even bother moving my lips or trying to appear involved. Dwight must be daydreaming, too, because he isn't singing, either.

"Sang!" Mizz Pickett commands me and Dwight.

Snapping to attention, we sing—just like two unwilling slaves ordered to entertain the master on Christmas Eve.

"What's ill ya? Clap ya hands!"

We clap our hands.

She endures two seconds of that and lifts her right index finger to the sky, circles it around her head and points to the door, barking, "Yawl niggas get out. Walk home."

We had been sent to the back before to wait, behind the swinging saloon doors in the entry area. But this time she was making us walk the whole ten miles home, because we didn't sing and clap the right way. A pro at not letting Mizz Pickett see when she's getting to me, I am cool as I stand and follow Dwight down the aisle, out the swinging doors. Two outlaws getting run out of town.

Outside it's cool, too, but we're boiling mad once we hit the sidewalk, marching ourselves up Woodland Avenue, the two of us grumbling loud enough not to hear our stomachs reminding us that dinnertime is approaching. Besides the cookies, we're used to getting Sunday dinner after church. Sundays aren't like those other special-occasion dinners when the Picketts eat nocturnal animals. Sunday is basic chicken, rice, gravy, and mixed vegetables, plus Kool-Aid for Mr. Pickett.

"Don't talk about no food!" Dwight threatens.

I don't argue. Instead, we window-shop, pointing out different things we wish we could have in stores. We play the car game, seeing who can claim the coolest car that drives by as theirs. By the time we pass Mt. Sinai Hospital on 105 near Euclid Avenue, we're having fun, entertaining ourselves as best we can.

Over the years, I had passed by the hospital many times, but always as a passenger in a car or a bus. The hospital had an old wing and a new wing, making it an odd sight. The old part of the building was a two-story brownstone and white marble while the new part was ten stories, white and light blue in color, and built right onto the original wing. When we come close to the old wing, I stop and stare. Being so close to it, I realize for the first time that it's connected and still a functioning part of the hospital. Before now, I had assumed it was an abandoned building.

I'm not sure why it interests me so much, but I walk toward the old wing to get a closer look.

"C'mon," Dwight says, tugging at me, "it's gettin' dark."

We make it from the church to the house in a little over two hours. The station wagon is in the driveway and when we come in the side door and up the two steps into the kitchen, we see Mr. and Mizz Pickett sitting at the table, their plates heaped with food. We don't stop, but continue to the next set of stairs that take us up to the first landing. Mizz Pickett motions with her fork at us, telling Mr. Pickett, "Dhere dey go."

Stomping up those stairs, I let loose, "And I ain't goin' to church no more."

With my back to Mizz Pickett, I hear her hop out of her seat as she comes to the foot of the steps and hollers after me, "You don't wanna go, I ain't gone make you go. But you ain't gone sit here in my house, watchin' my TV and go in my 'fridgerator! Ya hear me, Boy?"

Saying nothing, I quicken my step up to the second landing, knowing good and well that I've made an announcement nobody's made before. Quiet descends on the house, as if everyone is in shock that painfully shy Twonny has spoken out like that. Dwight comes into the bedroom with me, his eyes as big as half-dollars. He hides his envy, but I can tell he wishes he had the guts to say it, too.

When the next Sunday came, I didn't get dressed to go to church. True to her word, Mizz Pickett locked me out of the house and made me wait on the porch. That Sunday and the next Sunday, I spent pleasant mornings sitting out on the porch, daydreaming, watching the birds fly from tree to tree. The following Sunday, it occurred to me that there was no one to enforce my sitting on the porch. From then on, every Sunday morning, after the Picketts' car had pulled out of sight, I was off on an adventure—four hours of freedom.

One of my favorite pastimes was climbing up on the mailbox at the corner of Parkwood and Primrose, right by Mr. Heywood's grocery store. Straddled high atop it, I could watch the comings and goings of the neighborhood as a king might watch his subjects. Inevitably, the local patrolman who drove the beat would show up and pull close to the curb, sometimes screeching his lime-green police car to a stop like he was going to jump out and pull his gun to arrest me. Every week he'd yell, "Get off that damn mailbox, I told you!" I'd jump off and as soon as he left, climb back on and sit there waiting for him to return. Once I even stood up on top of the mailbox to really annoy him. We got to have an entertaining routine going.

Mizz Pickett never knew about any of that. She was too busy finding every opportunity to remind me that I was a heathen. Sometimes she was surprisingly subtle. She'd walk behind me at dinner after they'd been to church and sing, "I-I-I, lo-o-ve the La-a-a-awaw-ed," with a falsetto trill, saying to her husband, "We sho' did have a time in the Lawd today, didn't we, Day-di?"

"Sho'nuf."

Sho'nuf? I'd think in my mind, talking back—well, I had a good time in the neighborhood, jumping off the porch, playing, running up and down the streets, sitting on a mailbox, and teasing the policeman.

But showing no emotion at all, I'd stand and ask to be excused in the usual way, saying, "Thank you, Mu-deah. That was good."

She'd wave me off in her usual way, with a look of complete disdain.

Subtle or unsubtle disapproval about the stance I'd taken, it didn't matter. And so, while my spiritual education was far from over, from that time on I never went to church again. Other terrors remained in my life. But as for those ghost stories, I just wasn't going to listen to them anymore.

Want You Back" was the song that put a bookmark in time, 1970, just following our move to the new neighborhood, when I was ten. The song meant everything to me—singing, dancing, pretending I could sing and dance like Michael (only when I was alone); and wanting to dress and look like the Jackson 5, with their fringe vests, bell-bottoms, and various-sized afros; and, of course, what I yearned for all the time—to hear them sing about being in love with some pretty, pretty, sweet girl. As music was doing for me more and more, "I Want You Back" gave me sanctuary, a place to go with all my romance, longing, sadness, and dreams. Words and melodies helped me travel in my mind the way light and image had before; music helped me connect to that higher power and made me feel safe. Music was my refuge, a place for me to lay my burdens down, and the singers were the preachers and teachers I cared to hear.

Technically, the Picketts didn't allow secular music at home. The Devil's music, they called it. Sounds of scratchy LPs with the voices of the great gospel singers of the day—James Cleveland, Shirley Caesar, the Five Blind Boys of Alabama, and the Mighty Clouds of Joy—filled the house most of the time. But when Mercy and Lizzie were around, they played their own records or the radio or tuned the television to the music shows like *American Bandstand.*

I also heard the latest music pouring out of the mom-and-pop record stores on 105 and over at Manhattan, the barbershop at 105 and St. Clair, next door to the Rexall's. At Manhattan, I loved to sit and listen to the music playing underneath the storytelling of the older men and watch the younger men that came through, looking cool in their hip clothes and two-tone platform shoes, and making me laugh when they sat under hair dryers like the ladies.

The barbershop was farther down the street from the Picketts' thrift store, the one they opened up around the time of the move to the new house. After school, on weekends, and in the summer when I wasn't in summer school or going out to work with Mr. Pickett in his landscaping business, I was put to work at the thrift shop. There, on warm days, I could enjoy the music blaring from the record shop directly across the street while I worked.

One flaming hot afternoon in July 1970, I arrived about fifteen minutes early for my appointed time to work and spotted Mizz Pickett just inside the open door. At first, I couldn't believe my eyes. It had to be a mirage: Mizz Pickett dancing by herself inside the store to the pounding music coming from the speakers set up outside across the street. It was the Temptations song "Ball of Confusion." She was lost in the rhythm and heat, shaking her hips back and forth, making up her own footwork while her eyes closed with her cat-eyed glasses almost falling off her face, as she moved her mouth to the words: *"Ball of confusion! That's what the world is today. . . ."* This was no mirage, which I knew the instant she opened up her eyes and caught me standing there, frozen, with delighted shock written on my face. I had busted her dancing to the Devil's music.

Mizz Pickett scowled as she cut her last step and swatted at me, yelling, "Get on in here!"

The thrift shop occupied a large space, possibly what had once been a small department store, and was equipped with a delivery ramp in back. The place was rat-infested and jammed with clothes that no one, at least no one I knew, wanted—dresses, suits, and shoes from earlier eras, ballroom gowns, tuxedos, wedding dresses from the thirties, and hundreds of hangers everywhere. There was an abundance of pipes from the hardware store, in straight and L shapes, lying around, in case they were needed for hanging more clothes on. It was musty smelling in the store, along with the other damp, questionable smells brought in by all those clothes. Used suitcases, hat boxes, and old steamer trunks were stacked on the floor. There were costumes from different time periods like burlesque shows—feather dresses and boas—and jewelry of all sorts, plus old-fashioned pairs of glasses, even an occasional set of false teeth. Scarred mannequins, many of them missing limbs, stood or leaned in miscellaneous spots in the store, among a variety of different furniture and household items, small and large—everything from toasters and lamps to refrigerators. Covering a few counters were lots of old electronic appliances, televisions from the very early sixties, radios, and record players, some of which worked and some of which didn't. Enchanted by the glowing tubes I saw in their intestines, I liked to tinker around, exchanging tubes between TVs and plugging them into others. Through dumb luck, I managed to fix a few. When I did, Mizz Pickett declared that I was an electronics genius and dragged in all the broken television sets she could find for me to repair. Sometimes I was successful at it, but the fact that I knew nothing about voltage—or the potential hazards of messing around with electrical wiring—made it a miracle that I didn't blow up the whole place.

My triumph at the thrift shop was the day I offered to do window-dressing. Remembering the store windows I'd seen downtown on my outings with Keith and his mother, I got the idea to create a wedding display in the window with a few mannequins, wedding dresses, and tuxedos. When I finished, Mizz Pickett was elated at how it transformed the ambience of the store, giving the rundown graveyard for other people's tossed-off things a sort of quaint, antique appeal.

Window-dressing soon became my special task, after the regular duties I shared with Dwight of cleaning whatever items had recently arrived. By this time, cleaning came easily to me. Maybe because it was the only way I had to create order in a life that was otherwise my own ball of confusion; or maybe because it was one of the only ways I had to win approval from Mu-deah. Whatever the reason, I savored the rare bones of praise she threw me and tried to do a thorough job.

I had a tough challenge with an ancient pair of Stacy Adams men's wing tips that were so wrinkled and curled up at the toes that no amount of polish and newspaper stuffed down in them improved their weathered state. Finally, I was able to find wooden shoe trees to put in them that straightened out the old shoes. It seemed you could know the life of the person, all his trouble and unhappiness, by his shoes, and I couldn't stop wondering about the man who had worn these wing tips. He had to be the most sorrowful person in the world, I thought, to have walked so many miles to make his shoes curl up so. Nobody should have to walk that much, I thought, and I felt sad for the former owner of those shoes for the rest of that day.

Mizz Pickett needed more manpower than what Dwight and I could offer, so she hired a fellow by the name of Frank, a vagrant who was traveling through Cleveland and who must have fancied himself as English royalty. Frank wore a polka-dot ascot, a broken-down fedora, and pants that only went to his shins—making him look funny with his dress shoes and pinstriped silk socks. Every day he would politely ask Mizz Pickett, "May I bother you to give me a dime for a cup of tea?"

I had no idea where he was going to get a cup of tea on 105. Most of the guys who looked like him were chugging Thunderbird and Mad Dog, the street name for Mogen David 20-20 wine.

Dwight and I were at the store the day Mizz Pickett discovered that somebody had been urinating in the deep sink in back. After a thorough investigation, she figured out that the culprit was Frank, and she lit into him with as much disgust as if it had been one of us.

Frank shrugged, smoothed his ascot, and replied, "Well, Miss Pickett, sometimes it's a must."

For weeks, Dwight and I laughed about that in private, using the line "Sometimes it's a must" in every context we could. Meanwhile, Mizz Pickett appropriated the phrase for her own *edification,* to use her favorite high-powered word.

Even though the work we had to do for Mr. Pickett was physically harder than working at the thrift store, I preferred being with him to Mizz Pickett any day. Knowing so little about the man I called Dada, I thought the contact would allow me to know him better. But away from his Sunday pulpit, Ulysses Pickett hardly said three words.

I learned something about him one murderous hot day, the air so humid it was hard to breathe, when we were doing yard work out in the suburbs, where all the upper-middle-class white people lived. After many hours, the lady of the house, a young, vivacious woman, came outside on her porch. She was wearing a light summer dress and the wind pushed against her body, revealing the outline of her naked form. Holding a tray with glasses and bottles of cold 7UP on it, she called to Mr. Pickett. Dwight and I followed behind him as he approached the porch.

"I thought you and your boys might be thirsty," the woman said, extending the tray.

As we took the bottles, Mr. Pickett mumbled, "Thank you, Miss Ma'am."

"You and the boys are quite welcome," she replied.

The entire time Mr. Pickett kept his head down and his eyes lowered—as if a Klansman's voice was screaming in his mind, *"Niggerrr!! Don't you look!"* So he never once looked up at the white lady. It was strange for me to see, since I was the one who kept his head down around everyone. And that subservient pose of his was in such contrast to the powerful man I knew him to be that it confused and troubled me for some time. It made me finally understand why he kept a jar in the back of the pickup for peeing in, because he didn't want to ask to go inside people's homes to use their bathrooms.

It hurt me to know of Reverend Pickett's shame, so I translated it into more of my own.

Though my only rebellion so far has been in refusing to go to church, I make the decision, falling to sleep in tears one Friday night after hours of being berated, that I'm going to steal away for a day and stay as far away from the house as possible.

On Saturday, I rise earlier than even Mr. and Mizz Pickett, who usually get up with the birds, and slip out of the house without being noticed. For most of the morning I walk around the neighborhood, passing houses of friends and classmates. Soon the smell of breakfast fills the street: pork sausage coming from Jessie's house, eggs and toast from Michael's house. Before long, they're up, too, and we're off to have fun for the rest of the day. Jessie has so many brothers and sisters, his mother never seems to notice that I'm there. Another mouth to feed? She has to cook up so much food anyway, she just fills my plate as if I live at her house.

"Oooooh, you're gonna get it," Dwight says as I try to sneak back into the kitchen that night around nine o'clock. Standing at her brother's side, Flo adds, "Mudeah's been looking for you."

"What'd I do?" I ask.

They go on together about how they've spent the day

cleaning the house without me and that Mizz Pickett has been mumbling the whole time how she is going to wring my neck when I get home.

All the fun freedom of my quiet uprising drains from my memory as I become justifiably scared.

"Twonny!" Mizz Pickett roars from upstairs. "Is that that nigga?" Then I hear her running down the steps, linoleum-covered just like the ones at the old house, till she makes it to the lower landing. The sound of her making the turn to go down the second flight of stairs is like a dog's claws slipping, sliding, and scratching as if trying to grip the linoleum.

The next thing we hear is her falling down the second flight of stairs. The three of us rush over to the foot of the steps, getting there before she even comes into view. To me it is such a sweet sound—Mizz Pickett falling down those slippery stairs. It's the music of a delightful song that almost makes me want to dance, all but forgetting the trouble I'm in.

Next I see Mizz Pickett cascading down toward us. She is wearing only one shoe as she crashes at the bottom of the stairs, grabbing at the air and hollering with loud shrieks. What a sight—like she just got a good spanking.

Taking what could possibly be her last breath in this world, she looks up at me. Shaken by the fall, with her glasses sitting on her mouth, she reaches out to me with two hands, clawing the air, and says in her absolute meanest voice, "Nigga, where you been?!"

Those words are the best she can muster. I watch in relief as she retreats back upstairs, rubbing her hips, thighs, and kneecaps.

The following Saturday as I'm outside raking leaves, thinking about other ways to avoid Mizz Pickett, I spot Keith walking fast up the sidewalk. He frequently comes over on the weekends, sometimes because he misses us, sometimes because his mother has to go somewhere and arranges with Mizz Pickett for him to stay. It always cheers me to see my handsome little brother Keith, even though his face often shows distress that I don't understand.

This is one of those Saturdays that Keith has walked over on his own without his mother calling and Mizz Pickett acts pleasantly surprised to see him. After my chores, Keith and I are upstairs playing when Mizz Pickett calls up from the kitchen to say that there's a phone call for Keith.

This is somewhat unusual, considering that Keith doesn't live with the Picketts anymore and if it was his mom, Mizz Pickett would have said it. Flo has friends from high school who call sometimes, but Dwight and I never get phone calls. Keith shows no emotion at all as he goes into Mizz Pickett's bedroom to take the call.

Five minutes later I hear Mizz Pickett calling me, "Twonny! Get Keith and bring him down here." She doesn't call me *nigga* for a change, which makes me hope I'm not in trouble for something, but when I go to get Keith, I can't find him anywhere. For some reason, I don't know why, he's hiding. At last I find him huddled in the corner of Mr. Pickett's closet.

"What'choo doin', Keith, you in trouble? What happened?"

Keith sits there, refusing to say a word or to budge.

Sure enough, when I go downstairs without him, Mizz Pickett says she wants Keith, not me, and to send him down.

After much cajoling, Keith goes downstairs. Over the next two weeks, I piece together why Mizz Pickett had to talk to him. According to Dwight, who heard it from Flo, who heard it from Mizz Pickett, there was a man on the phone asking for Keith. Suspicious, Mizz Pickett listened in when Keith took the call and heard the man ask if Keith had been telling anyone what they'd been doing together. To this day, I applaud Mizz Pickett for listening in and trying to protect Keith.

Understanding what has happened, it is the first time I've ever known anyone to have a secret like mine. And so the next time that Keith comes over and Dwight starts teasing him about it, calling him sissy, I go after Dwight, ready to fight him with all my pent-up outrage about what had gone on in the Picketts' basement at the old house.

Since Dwight almost always initiates our fights, he is caught off guard, and knowing the trouble he'll be in if Mizz Pickett gets wind of what he's been saying, he stops teasing Keith.

It has been more than two years, now that I am ten, since Willenda has had an opportunity to get me alone, mainly because she hasn't been around much. Whenever she is around, I watch the movements of others, making sure never to let myself get stuck alone with her—not in the house, not even in a room, if I can help it.

In Willenda's eyes, there is no feeling. Nothing. In a cloud of Bel Air cigarette smoke, she can stare at her soap operas and at her news shows, completely engrossed but completely disconnected. I can't read her or her moods, or predict her behavior toward me, so I continue to keep my distance.

That's what I'm doing this Friday night when Reverend and Mizz Pickett have taken Dwight and Flo to a tent meeting church revival and left me in her care.

"Twonny," Willenda calls down to my bedroom from the top of the attic steps, "you take your bath?"

"Yes."

"No you haven't," she calls back. "I can smell you. Go run the bathwater, and I'll show you how to take your bath."

Before I have time to question it, my response jumps out of my mouth with a shout: "I don't need nobody to show me how to take no bath!"

There is a long pause. Willenda says, "Okay."

Just like that. It was over.

Not long after I began the fourth grade at Parkwood Elementary, I became known as Fish. When my classmates gave me the nickname—cut from my surname— I didn't like it; I preferred Antwone, especially because I rarely heard it. On the other hand, I couldn't stand being called Twonny, so Fish was something of a promotion. In fact, after a while I started to think it sounded sturdy, like a person with some personality.

With a new name and enrollment in a different school, I was given a chance to start over—to reinvent myself. And a big part of the new me was a woman by the name of Mrs. Profit, a teacher in the truest and best sense of the word. If there is such a thing as human beings who act as angels in our lives, Brenda Profit was that for me. And perhaps she played a similar role in the lives of many other students over the years. In any case, it would be some time before I really appreciated everything Mrs. Profit had done for me.

She comes into focus in my memory a few weeks after my first day at Parkwood Elementary. I see her that morning as she often looked when the weather was still warm outside, dressed in one of her clean, flowered-print cotton dresses, belted at the waist, the hem modestly at her knees. Standing at her desk as we file in, she is an attractive, brown-skinned young woman, slender, with styled hair cut just below her ears, groomed neatly in every detail.

Like me, Brenda Profit is new to Parkwood Elementary. With her husband, Milton, recently discharged from the army, she has come to Cleveland from New Orleans to make a new start in life and to build their dreams in this family-friendly city. She will later tell me how, as a little girl, she was fascinated by the Great Lakes and dreamed of visiting them one day. And here she is. Another dream was to have her own classroom. She got it. And we came with it—a package deal.

Accustomed as I am to reading people, what I already know about Mrs. Profit is that she is fair. She treats everyone as special, me included. Not just the prettier girls or the more handsome boys. She gives each of us the same chance to do well and to do those special helping jobs, like clapping the erasers. It's a glorious feeling I'll remember the rest of my life, the first day I got to go out on the fire escape and clap the erasers. The sweet autumn breezes blowing the chalk dust all around me, it seemed like a dream. But it was real. And it happened simply because I asked.

Mrs. Profit is as fair with discipline as she is with encouragement. If, for example, one of the kids who used to be the type to get away with misbehaving is acting up, he or she gets sent to the coatroom for five minutes just like the rest of us.

As we trickle into the classroom this particular morning, greeting Mrs. Profit one by one, we behold a strange sight. Instead of the rows of desks lined up from front to back, she has rearranged the classroom with the desks facing each other, side by side. This means that there is no longer a front row for the "better" kids or a back row for the "bad" ones. Now each desk faces someone else and is next to someone else. Not only that, but she has arranged the desks in girl-boy order, every girl seated next to a boy and across from a boy.

Mrs. Profit doesn't explain why she has done this, nor the many benefits that will arise. For me, with all my romantic yearnings cooped up inside my painfully shy shell, this is about to be a life-transforming event, even if I don't know it yet.

Suddenly, I don't have to sit and dream of a little girl who wafts by in a yellow dress and perfume on her way to the front row. Girls are sitting next to me, across from me, becoming my friends. My first new friend, seated to my right, is Janine King, who lives in my neighborhood. For two weeks, I worry that Janine will see inside my messy desk and embarrass me. So whenever I get something out of it, I lift the lid as fast as I can and grab whatever it is with lightning speed before shutting the lid.

Then one morning, to my chagrin, I arrive a few minutes after Janine and see that she has the lid of my desk lifted open as she gazes inside disapprovingly. Janine looks up at me and says, "Fish, no wonder you don't want me looking in here." For a minute, I feel the blood rush into my cheeks in shame. But before I can feel too terrible, Janine offers to show me how to straighten up my things. From this day on, I keep the inside of my desk shipshape.

Young as she was, Mrs. Profit really knew what she was doing. With constructive criticism, she encouraged rather than condemned. She found something to compliment in each of us—a neat paper, a good attitude, an eager face—and rewarded the whole class for our overall positive efforts with impromptu parties, field trips, and other celebrations. We became a family. And she shared Milton with us.

On Fridays, for example, when he comes to pick her up, just after the last bell of the day, we're still hanging around the classroom when he arrives and he greets us all with cordial hellos. Milton Profit is an average-looking black man, not super handsome or dashing, but a true gentleman. And when he sees his wife, he looks at her as if seeing her beauty for the first time, like he can't believe he is so lucky as to be married to her. They hug and kiss, two lovebirds. Not too much, but enough to show how much they cherish each other.

The Profits are the first married couple I have known to show believable affection, and Milton is the first husband I have seen being romantic and loving toward his wife. Theirs is a picture of domestic caring I will hold in my vision for years to come.

Something else that starts to affect me early on is my sense that Mrs. Profit is a woman of faith. It isn't anything she says specifically about religion or God; it's the few moments of silence she gives us every morning just after the breakfast program. She calls it a time to be quiet and lets us know that we can meditate or think about whatever is meaningful to us or to say a silent prayer. Even though I recognize a religious quality to this space of time, it is so different from the Picketts' hellfire and damnation that I don't feel any pressure. At first, when it's time to put my head down in silence, I still feel that I'm much too bad to talk directly to God, so I mentally recite the Lord's Prayer. Slowly, over time, I discover that I've become more conversant with God and that it's okay to talk to Him or Her like a friend, someone who maybe is looking out for me.

After a while, sometimes, in bed at night, I find myself opening up in a way I can't with any person, and I allow my tears to fall and let myself feel all the despair and anguish that living in the Picketts' home has compounded in my psyche. And in these private moments I ask God the most pressing question of my life as a ward of the state: When is the good part gonna come?

During my first year with Mrs. Profit as my teacher, the changes in me took place slowly and subtly. If self-esteem was what you used to fill up like a tank of gas, the Picketts had siphoned mine out to nothing. Mrs. Profit helped change all that. Bit by bit, a few drops were beginning to accumulate as I learned to like myself more than I had before.

Unfortunately, that wasn't yet in evidence when it came to my fourth grade academic performance. Aside from art class, in which I continued to excel and which I continued to love, the other subjects required that I speak up in class, which was close to impossible because of my oppressive shyness, not to mention my nervous stutter that the speech therapist called "lazy mouth." More than anything, it was the physical and psychological exhaustion from the toll being taken outside of school. At the end of the year, my grades were so poor that Mrs. Profit was put in the position of having to determine whether to hold me back and have me repeat the fourth grade again, or take a risk and promote me to the fifth grade. She took that risk, a small miracle in my favor, which coincided with an even greater miracle.

Toward the end of the year, a group of parents approached the principal of Parkwood Elementary and asked if the school would consider having Mrs. Profit stay with her class and be our teacher for the fifth grade. The school approved, Mrs. Profit agreed, and she did it again the following year. By deciding to mentor us from the fourth grade all the way through the sixth, she guided us on our difficult journey from childhood to adolescence, preparing us not only for our graduation from grade school to junior high, but actually preparing us for life.

The improvement becomes clear to me on a winter morning in 1971, in the middle of the fifth grade. Eleven and a half years old, I'm sitting in a reading circle with Mrs. Profit at our head and we are going around the circle, each student reading a page out loud. When it gets to be my turn, instead of giving in to panic, I calmly and carefully read my page without error—including one long, unfamiliar word.

"How did you know that word?" Mrs. Profit asks, stopping the next person from reading so she can talk to me.

"I just broke it down," I say softly, "how you told me to."

Mrs. Profit smiles and nods, pausing before she says, "Antwone, I'm proud of you. I want you to know that I really struggled over promoting you, and I'm so glad that I did." She smiles again and comments, "You are doing very well this year."

Her honest, careful words are the equivalent of lightning bolts and thunderclaps. Outside I shyly accept her praise, but inside I'm flying with the birth of revelation. It's the first time I've ever realized that there is something I can do to make things different for myself. Not just me, but anyone. That no matter how often someone says you can't do something, by simply working harder and trying, you can prove them wrong and actually change your circumstance. This lesson is a piece of gold I'll keep tucked in my back pocket the rest of my life.

A highlight of the fifth grade is when we are taken on our first field trip to hear the Cleveland Orchestra. Janiece Womack, a pretty brown-skinned girl who lives two blocks from my house on Parkwood, and her sister,

Daniece, come to school dressed alike, each in dresses with sequins and a big brown bow. Mrs. Profit is so tickled that she brings Janiece out in the hall with her sister and her sister's teacher as everyone admires the Womack girls. By this time, because of Mrs. Profit's fairness, Janiece's compliment is my compliment, just as mine is hers and everyone else's.

Outside the concert hall, buses from schools all around the city unload nicely dressed children from all the different neighborhoods—black, white, rich, and poor. As we all flock inside, there is a certain awe to this event, almost like going to a big church. Some of my friends try to say the classical music is corny and I pretend to agree. But deep down I'm interested, even eager, to hear it.

Inside the great hall, Mrs. Profit leads us to our seats. Her dignified expression tells me that we are somewhere important and we deserve to be. On the stage, musicians are tuning up their instruments, some playing scales, some fragments, like different-voiced creatures talking to each other in languages the others don't understand. This cast of thousands reminds me of watching the old MGM movie musicals—all the men in tuxedos and the ladies in long dresses, with lots of fanfare and applause when the maestro comes onstage. He identifies the various instruments and we get to hear how they sound alone, some of them, like the harpsichord, that I've never heard before. Then he waves his baton for silence and holds it in the air. All sound and motion cease in perfect obedience. And suddenly he points his baton and begins waving it wildly, with his other hand pointing and moving, too, as the music begins—filling my ears with an exciting melody made up of lots of smaller melodies.

Like the songs on the records I love that carry me away, this music transports me out of time and place. In my mind, I imagine standing inside the music itself, wrapping myself in the melodies of the individual instruments. I imagine standing next to the cello, hearing it alone, and next to the French horn, and the big kettle-drum. As I continue to listen, I realize there is more to music than the standard blues chords I'm used to. This is much more intricate, with everyone playing his or her own different part. But it all goes together. Amazing. Then my ear keys in to the sound of the triangle that comes every now and then at the end of a long phrase. It's a single note, a small *ringggg* of punctuation. I wonder what the music would sound like without the triangle. That's when I understand that it wouldn't sound complete and that in this orchestra, everything has its place, and even the smallest of the instruments is important.

Toward the end of the fifth grade, Mrs. Profit escorts us on a weeklong school field trip to Camp Hiram House, a wooded estate in the Cleveland area. There we stay in cabins at night and study science and history in the out-of-doors during the day. Together with counselors who are black and white, we reenact the story of the Underground Railroad. Every night before we go to sleep, Mrs. Profit comes into our cabin where we're dressed in our pj's, boys on one side of the room and girls on the other. She checks to make sure we're settled in and comfortable. Before she leaves us each evening, Mrs. Profit leans over to hug and kiss a few of the students good night. Even though I'm not among the fortunate ones, I feel each night as if she has personally tucked me in and that I am safe.

From this time on, Brenda Profit became more than a teacher to me; in fact, she was the woman closest to a mother that I would ever know.

On my way to adolescence, I was now in dire need of

a parental figure to explain some of the physical changes that were starting to happen to me. Mizz Pickett, I knew, couldn't be counted on. A few years before, when Flo began her menstruation, instead of explaining that it was normal and what to do for herself, Mizz Pickett made disgusted noises whenever Flo came in the room. *"Hun-uhn,"* she grunted once, saying to Flo, "La-a-a-wd have mercy! Heifer, did you bathe?"

Sometime later I found out that a heifer was a cow. Mizz Pickett was always comparing us to animals—like polecats and road lizards. She used the word *shee-ann* about Flo, something I later heard was the smell of wet sheep's wool.

Then Dwight hit puberty and was horrified to go through a phase when his nipples started to hurt like crazy and his smell began to change. No one ever told him that it was also normal for boys to go through something like this—so he was convinced he was turning into a woman. Scared, angry, and in terrible pain, there was nothing he could do to get any relief. And as for the smell of his increased perspiration, Mizz Pickett's solution was to shoo him outside, like you would a mangy dog. That's how they treated Dwight, even worse than Laddie, the Picketts' dog. No wonder Dwight got so mean; they made him so.

And the tragedy of that was how naturally brilliant he was. He didn't just read everything he could get his hands on—books, magazines, newspapers, cereal packaging—he also managed to retain all the information. Dwight was the type of person who could have grown up to be a senator or a congressman or a CEO, somebody really important and respected, if only someone had loved him and given him the right encouragement.

After Dwight went through it, I started to have the

same pain in my nipples. Knowing that it would pass, I wasn't as scared—even though I hurt. When it came to my underarm perspiration, Mizz Pickett didn't explain anything to me, either, but she did have a treatment for me—Clorox. She would pour liquid bleach into the bathwater and make me scrub myself with the suds. So now I didn't smell like an adolescent boy, I smelled like a wash gallon of Clorox bleach.

As if in answer to prayers I hadn't even prayed, Mrs. Profit started addressing many of my concerns in school. She devoted a portion of health class not only to those more specific explanations of where babies came from, but also to general things like how to take a bath and groom oneself. The fact that she talked about things that related directly to me made me pay more attention to the other subjects that were still very foreign.

Mrs. Profit was so helpful that by the sixth grade, not only had my grades improved significantly, but I had also developed something of a sense of style, in spite of my severely limited resources. My afro, for example, became a real source of pride. I kept strawberry-scented Afro Sheen in my desk, along with my afro pick, and took them with me during my several trips to the bathroom every day. Mrs. Profit never questioned the frequency of my bathroom visits. She probably didn't have to, not with the thick trail of strawberry aroma that must have wafted back into the classroom upon my return each time.

Early in the school year, Mrs. Profit took us on a field trip to Cleveland's Space Museum, and as we were being shown the various electronic and futuristic devices, the program director announced that one student would be chosen to appear on the video television monitor. Before I could choke from surprise, that one student chosen, from all the students out of all the schools who came on the field trip, turned out to be me. I got to go in front of the camera, then turn with the rest of the kids to see me, not a bad-looking guy, right on the TV. I was dressed in my blue jeans and blue long-sleeved sweater with a deeper blue stripe down the arms, in my medium to full afro, smiling like any normal, happy twelve-year-old boy.

Nothing made me more proud, until the day came for our next field trip—which was to the Cleveland Health Museum. There our class joined with several other classes as we were shown through the many exhibits. At the exhibit showing the reproductive process, the program director pointed out the different stages of the baby's development, asking all of us, "And who knows what the liquid is called that protects the fetus in the mother's womb?"

My hand shot up first and the program director pointed to me.

"Amniotic fluid," I said.

"Very good," said the program director, looking impressed.

I glanced quickly at Mrs. Profit, and she was beaming.

My self-image was improving, thanks to Mrs. Profit, my classmates, and to some of the changing social attitudes. Fortunately, black pride had made its way to Glenville in the early years of the seventies. Up and down 105 many of the stores showed posters of some of the more famous images of the movement—Angela Davis, Stokely Carmichael, and pictures of the black power fist raised high by unnamed black men and women. Seeing that the community was behind these activists helped me gain greater acceptance of my own dark skin, helping to undo some of the damage Lizzie Pickett had wreaked by calling me derogatory names that referred to my color.

For her part, Mizz Pickett didn't seem one bit pleased

about any of the political black organizations. She constantly worried that the Black Panthers and the other more militant groups were going to be the ruination of all God-fearing Negroes. Of course, whenever we passed any of the guys on the street who were known to be active in any of these organizations, she'd give them the black power fist in solidarity. Meanwhile, she was muttering under her breath, "You trouble-makin' niggas!" And then she'd add, "They know me. They ain't gonna mess wit' me," never realizing that they weren't the enemy to begin with.

These groups, combined with the headlines about the Patty Hearst kidnapping and mail bombers, push Mizz Pickett over the edge this one day when I'm riding in the car with her, coming home from the thrift shop. She's taunting me as a heathen, calling me names like Frank, Jr., suggesting, I suppose, that my refusal to attend church will turn me into a vagrant with delusions of grandeur like him. More and more, Mizz Pickett has substituted psychological beatings for physical ones, knowing as she does that they bother me so much more at this point.

It's a warm day and all the windows are full open. I'm sitting in the passenger seat, leaning my head out, trying to ignore her and catch whatever breeze I can on my face. In between us on the wide plastic-covered seat is a VHF antenna that belongs to Lizzie. Mizz Pickett appears to notice it for the first time, without any recognition of what it is or how it came to be here. She pulls her glasses all the way off her face, looks as close at it as she can, real fast, and then gets my attention. "Nigga!" she says, bugging her eyes out. "A bomb!"

With that, she slings the entire VHF antenna out the window into the middle of the street and speeds off. Two blocks later she comes to her senses, screeches to a stop, pulls over to the side of the street, and makes me get out to run back and retrieve it.

This event coincided with the realization that on top of being cruel, Mizz Pickett was probably crazy. But instead of freeing me from the poison arrows of her words that told me I was worthless and would always be worthless, the recognition of what I took to be her insanity only plunged me further into despair. With many of my old defenses weakening, I worried that it was only a matter of time before they turned me as mean and crazy as them. These were the nights I cried myself to sleep, asking God why—why I had to live with the Picketts, why I couldn't know my parents, why it was taking so long for the good part to come.

By day, my life with Mrs. Profit and the rest of my family at school gave me breathing room, as did a new defense I was beginning to cultivate—a sense of humor. In fact, I found out that when I started telling my friends stories about Mizz Pickett, like the day she threw the antenna out of the window, or when I would just imitate the way she talked and what she said, they would laugh. Even Mizz Pickett's grandsons, Teresa's boys, laughed at my impersonations.

None of the adult Picketts ever picked up that I was secretly parodying Mu-deah, nor that I had the ability to be funny, except maybe for Mizz Pickett's nephew, who was living with us at the time. Brother, that's what they all called him, and so did we. He was in his mid-forties, a man who wasn't heavy and wasn't thin, wasn't good-looking and wasn't ugly. He may have once been more handsome, but life had drained him of any noticeable physical appeal. Brother had come up from Mississippi to get a rest in Cleveland after his marriage broke up. I didn't know the whole story about Brother, but it was obvious that he was brokenhearted and in need of a lot of cheering up.

In this time period, Mizz Pickett and Reverend Pickett were almost always gone during the day, if not for a few days' stretch here and there. Mizz Pickett was starting to make trips down south to visit her ailing sister, and Mr. Pickett would take off regularly to go to out-of-town weekend church revivals. On these various occasions, since Brother was staying at the house, we were left in his care.

As depressed as he was, I caught on one day that I could make him laugh. We're in the den, me, Brother, Flo, and Dwight, and there's a cool, up-tempo song on the radio. I look at Brother to get his attention, then back at my fingers on the armchair, pretending to be surprised that they're tapping. Then I start wiggling them and my toes get in on the act. Now Dwight is grinning and Brother lets go a half smile. Like I can't help it, my legs dance on their own, pulling me up out of my seat, while my arms and torso begin to shimmy, until I'm dancing away to the music like a wild man. Dwight's howling, Flo is giggling, and Brother is actually chuckling, with an expression of real enjoyment on his face.

After that, Brother wants me to do my dancing routine whenever there's a chance, asking me, "C'mon, do that thing, you know, that you do." After my dance is over, he usually appears to be in a better mood and then asks Dwight to do his thing, which is to talk about all the interesting stuff he's been reading, like from *National Geographic* about animals and space exploration. Basically, I'm the entertainment, and Dwight is the education.

This continues for a while until one evening at the dinner table, Mizz Pickett starts complaining that Brother's been telling her about these special talents that Dwight and I have. She's acting mad, as if we've hidden our gifts from her on purpose. "Dwight, how come you don't tell me 'bout no Bengal tigers?" she taunts. "And you," she points her index finger at me, "I hear ya gotta little personality?"

I sink down in my chair, mortified with embarrassment for her to know I was dancing and had enjoyed it.

Looking uncomfortable, Brother hunches his shoulders, hands in his pockets, and gets up from the table, heading over to the radiator. He's cold all the time and stands there bending his knees back and forth in a fidgeting way. I can see he's been trying to tell Mizz Pickett that we're not bad kids, but when she starts up now about what hardheaded, rotten niggas we are, he doesn't argue. Brother's like that, political—nice to Dwight and me, but never standing up for us when we're being accused of doing things we didn't do or being treated unfairly. Sometimes, he'll come to me later, saying, "I don't mean no harm, you know how Mu-deah is. . . ." Though I act like it's no big deal, I sense that Brother knows I don't forgive him, or that I can't trust him.

Often Brother takes to the bedroom upstairs adjacent to ours for most of the day. The room has a certain smell, like decay. It comes from him, not because he doesn't bathe, but maybe just the smell of him getting older and the things he eats and drinks. He's not supposed to drink, but sometimes he does. Or maybe it's what the medications he takes do to him. He doesn't talk about that, but I know he takes them because the top of the chest of drawers is full of different pill bottles. Sometimes, after resting, he'll come downstairs and wander outside, leaving us alone for hours.

On one such sunny Saturday afternoon, Dwight and I are downstairs in the den, listening to the radio. Normally we'd be out with our respective friends on a pretty fall day like this one, but for some reason we're here—lying on the den floor, enjoying being in the house all alone. This is a popular radio station that plays only soul music, mainly up-tempo songs this particular afternoon. We're lying there getting into the beat of a record that's melting into the next one just as the voice of the black disc jockey bursts in with his rich, smooth, velvet voice almost shouting on top of the musical overture: "WJMO, *super* 14-90 on your radio. This is new and *super* bad by Jean Knight . . . 'Mr. Big Stuff'!"

At first, Dwight and I listen quietly to the lyrics, but in no time begin to sing along with gusto. Pretty soon, we make up our own lyrics: *"Mr. Big Dick, tell me, who do you think you are? Mr. Big Dick, you never gonna get my stank. . . ."* We sing at the top of our voices. We sing on and on, cracking each other up, using language that, of course, we're not supposed to be using.

Loud, uncontrolled laughter that is not our own interrupts us from the doorway. Scared and startled, Dwight and I immediately stop singing and turn to see Brother laughing uproariously, holding his sides. Panic swarms into my brain, but then I watch as Brother turns without saying a word to us and heads up the stairs, continuing to laugh like a mad scientist.

Dwight and I exchange momentarily relieved expressions, even though we know that he'll probably tell Mizz Pickett as soon as she gets home. When she does, we wait and wait, watching for a clue that we're about to get it. Dinner comes and there have been no explosions yet, so we surmise that we may well be in the clear. That is, until I look over at Brother and see him sitting at his place, barely touching his food, just looking at me with a big frightening grin. Frightening because he is clearly on the edge of laughing again.

I keep my eyes trained on him, silently begging him to keep his composure.

Mizz Pickett, noticing the disappearance of his otherwise sad face, comments, "It sho' is niiice to see ya smilin', Brother."

Now Dwight looks up at him and flashes nervous eyes, too.

Finally Brother can't take it anymore, not with all of us focused on him, and he explodes into laughter, laughing harder and in a more sinister way than he had earlier.

Reverend Pickett takes a deep swig of his beer stein of Kool-Aid and shoots a look at Mizz Pickett, as if to say, Did I miss something funny here or should you be calling the nuthouse (as she refers to it)? She sends a questioning look right at us, but then lets it pass. She even laughs along with Brother, like she's in on his private joke. I'm not looking at Dwight, but I know he's as grateful as I am that Brother didn't tell on us.

Besides these moments of escape through laughter and stolen time sketching alone in my room, music continued to be a release valve. Love songs let me dream of the day I could have a relationship like the ones they sang about. The Dramatics reigned supreme with a barrage of favorites, along with songs by the Temptations and Smokey Robinson and the duets by Tammi Terrell and Marvin Gaye.

But it wasn't Marvin's lead into the romantic I followed so much as his lead into sadness, which took me all the way down his road when he came out with *What's Going On?* Songs like "Save the Children" and "Mercy, Mercy Me" made me feel I understood him and feel that he would understand me, the way a father would. I loved Marvin Gaye, not as his personal fan, but as his personal son. That picture on the cover of the album told me everything I needed to know about him—a lone, handsome figure in shadows, standing in a shirt and tie and trench coat in the hard rain. I could see myself standing with him in the rain, dressed like him, sharing our sadness together. He could be my father not only because of all that we shared emotionally and spiritually, but also because he looked tall, as if he could look out for me in the world and not let anything bad happen.

With long gaps of time between records, Marvin Gaye tended to appear in my life and then disappear at various stages. This may have been the case for the rest of his public, but I doubt many took it as personally as I did when Marvin would retreat into artistic hibernation and then reemerge with something new. When he disappeared after 1971's *What's Going On?* I was concerned, and then when he came out toward the end of 1973 with *Let's Get It On,* I took every song as seriously as his earlier social ballads. Naive as I was, I had no idea that the entire album was about sexuality.

Maybe the topic of sex was still too tainted because of Willenda, or maybe it was because, unlike most adolescents my age, I was too distracted with other issues of basic life survival. Whatever the cause, I didn't get it yet that romance and sex were connected. Then again, I wasn't so sheltered that I hadn't heard sexually related curse words and jokes, like the lyrics Dwight and I made up that caused Brother to laugh so. There was one time that I used one of the really terrible words on Dwight after a long day of him tormenting me about my latest crush, Fatima Doyle. She was so cute and the best-dressed girl of my youth. She didn't just wear trendy clothes, they were of excellent quality, fit, and styling, not to mention so clean that her parents must have been paying for a lot of dry cleaning.

"When you-all gettin' married, Twonny?" Dwight

sneered. "Why you so in love with Fatima, man?" He was laying it on thick, and I was about to hit my boiling point.

As fate would have it, this was on one of the few days of late that Mizz Pickett had been favoring Dwight. When we entered the kitchen where she was at the stove cooking, she didn't hear him threaten me, "I'm goin' out to tell Fatima that you're in love with her," just before he rushed past her toward the door.

I tore after him, but before either of us made it out, Mizz Pickett grabbed me by the collar.

Dwight whirled around and presented himself as the plaintiff. "Mu-deah, Twonny keeps bothering me," he said, arms folded.

Seeing the anger on my face, she pulled my collar tighter and said, "You leave 'im alone. Ya hear me, Boy? Just let 'im alone and gone 'bout ya own bizness."

I gave Dwight my meanest look and he retaliated with the smirk he'd designed specifically to torture me.

"Motherfucker!" I yelled.

Dwight's smirk widened at the very instant I realized what I had just said and that I had said it in front of Mizz Pickett.

In more shock than me, Mizz Pickett stood there gaping and blinking behind her cat-eyed glasses for a lo-oo-oo-oo-ng moment. "Dwight, go-on outside. Twonny, get up to yo' room," she ordered us, saying that she would deal with me later.

In my room, I waited and worried about what pains I would endure for calling him "motherfucker"—something that brought me no real satisfaction, especially given what I would be made to suffer for saying it. When Mizz Pickett didn't come up, I became convinced that she was waiting for Mr. Pickett, the big guns, to come home and give it to me. After dinnertime came and went with me not being called down to eat, I figured she was letting me off easy with starvation. Then I heard her calling, "Twonny, get on down here and eat."

"Yes, ma'am," I said, hurrying down the steps and into the kitchen, where my place was set with a plate of food already.

Mizz Pickett was seated at one end of the long table and Mr. Pickett was at the other. Like stepping into a courtroom, I timidly took my seat in between the two of them and began eating, feeling all eyes upon me.

I was almost finished when Mizz Pickett broke the silence. "Who taught you dem words like dat?" When I didn't answer fast enough, she added, "You hear me, young'un?"

"Dwight," I lied with quiet resignation—a brilliant stroke of negotiating, I might point out, that put the blame squarely on him.

Mizz Pickett shot a sideways look at Mr. Pickett. "Mmm-mmm," she said. "That was what I thought. Up at that schoolhouse. Them bad young'uns, teachin' 'im all the no-good things of this and that and what-not. And bringin' it here to dis here house."

Mr. Pickett said, "Mmm-mmm."

I won my day in court and took it on home from there, saying wistfully, "Yes, ma'am, that's why I don't want to go to FDR next year."

Mizz Pickett didn't even hear. She was too busy hollering up to Dwight, "Nigga, you get down here!"

Of course, my knowledge of sex was not completely limited to curse words, nor was I completely disinterested in the subject. In fact, during Mrs. Profit's math class one day, I drew an elaborate portrait of a naked man with an erection together with a naked woman. With the notebook page turned one way, the man and woman were facing each other. When you turned it sideways, the border I'd drawn under the woman transformed itself into a bed, so now the man (with streaks of speed flying off him like the Swift Cleaners logo) was coming down on the woman.

"Fish," whispered Gary Spencer, the boy sitting the closest to me, "lemme see."

Casually and covertly, I slid the picture over to him facedown.

Gary scanned the room for spies and seeing none, flipped the page over. He smirked. Then I said, under my breath, "That's not all," and turned it sideways. Gary's jaw fell open, incredulous at my talents.

I sat there so smug and bad, until I felt Mrs. Profit's presence materializing right behind me. She cleared her throat, took the drawing from Gary, and began a familiar lecture, "Antwone . . ." I was sure she was going to make me write something like *I will not draw dirty pictures in class* one hundred times. She'd made me write the same sentence one hundred times so frequently, I even had a page ready with *I will not* already written one hundred times and a blank left for whatever misdeed I was never going to do again to be filled in. But Mrs. Profit was hip to me. Instead of that, she announced, "I want you to write the week's vocabulary words one hundred times."

"Yes, Mrs. Profit," I replied dutifully, and began the boring task of writing all twenty-five vocabulary words for the week times one hundred. Once again, she knew what she was doing. That week, for the first time ever, I got every single vocabulary word right on my spelling test, thus allowing me to feel pride instead of shame.

It was incidents like those that made me believe in my own possibilities. Toward the end of my time with Mrs. Profit there was a day when we stood together to sing "What the World Needs Now Is Love," and she asked each student to say what we wanted to be when we grew up. When it was my turn, I stood and said, "I want to be the greatest artist since Michelangelo." She had shown us pictures of the works of many world-renowned artists on the overhead projector, and he was my favorite. Plus, I wanted to answer in a way that pleased Mrs. Profit, and that really did.

How much Mrs. Profit knew about my status as a foster child, I wasn't sure. There was some contact between social services and the school administration, I knew, but for the rest I tried to distance myself as far as I could from the Picketts. The one time I ever mentioned anything to Mrs. Profit was on the last week of school when Mizz Pickett made me cut my afro for graduation. When Mrs. Profit noticed, asking me why I cut it, I told her the truth. She said nothing, but I felt she was disappointed and sad for me.

Later, in junior high when I grew my hair back, I wanted to let her see me. Not wanting to go empty-handed, I made a wooden paddle in shop class that said "Hot Hands" on one side and "Hot Pants" on the other, and took it as a present. It was a funny choice, as I came to see later. At the same time, even if I lived a thousand years and a hundred lives, I don't think I could ever do or say enough to fully thank Mrs. Profit for what she gave me.

Though she was glad to see me and touched by the gift, going back to Parkwood was one of those experiences that teach you not to try to go home again. Everything looked so much smaller and even older than it had before. Instead of returning to the family, as I had dreamed, there were students sitting in her classroom whom I didn't know and who didn't know me.

Other than the comment she made about my haircut, there was one earlier time, in the wintry March of 1972, in the middle of sixth grade, when I was twelve and a half, that Mrs. Profit led me to feel she knew or was concerned about my circumstances. This is a rare day when my two worlds collide. In the morning before I leave the house, Mizz Pickett promises me a beating when I get home for something or other that I've done or not done. As it happens, this week is my turn to collect the trash after our breakfast break at school. For the past few days, I've been walking around smiling like a Cheshire cat, feeling good to be asked to do something. But today, my look must be pensive enough for Mrs. Profit to see that I'm under some kind of stress.

I do try to hide it most of the day, but as the clock winds down, a sick feeling of dread comes into my stomach the second I remember what's waiting upon my return to the Pickett house.

A half hour before the final bell, Mrs. Profit comes over to me and talks softly so that others around can't hear. "Anything wrong?"

"No."

"Is everything okay at home?"

"Yes."

She says, "Okay," and lets it go at that. But I know from the concern on her face that she is wiser than she is in a position to tell me. When I go to the coatroom to put on my winter gear, she gives me another long look, and I try to think how to bring myself to tell her about what I'm feeling. But where to begin, I don't know, so I forge on—going to get my coat, and walking slowly down the corridor, away from the love and light of her classroom.

Parkwood Elementary is so old, there are actual foot-prints worn into the wooden steps, one track going up and the other coming down, grooves tracked into the wood by the heels of thousands of kids' shoes. I head outside feeling about as worn and used up as the old building.

Outside in the cold, there is a fresh layer of new snow and many of my classmates are sliding around and laughing on the playground. Some kids laugh as they dive into a spontaneous snowball fight, others fall onto the ground and wave their arms, making snow angels.

A few weeks before, I'd been out playing with Michael Shields and William Howell, the three of us walking down the block, when the overwhelming urge to cry came over me so suddenly I knew there was nothing to do to stop it. It came like a flood of water from the back of my head, ready to dump bucketloads out of my eyes—like having the uncontrollable urge to go to the bathroom. I turned the other way and ran toward the Pickett house as the tears fell.

"Where you goin', Fish?" Michael called.

"I gotta go home," I choked out, my voice probably giving me away.

I think of that, this day, as I watch the other children full of the wonderment that new snow brings, delighted with snow angels and snowball fights. To me, these children my own age seem new somehow; if they were wearing new clothes, in fact, they would even shine.

I don't feel new. My body feels old, weathered before its time, like tires on a car that's been driven too much in too short a period. And I think that even if I ever wore new clothes, I would still look dull or faded. Children are supposed to be new and most of them are. But not me.

*   *   *

154

A few weeks later, early spring 1972, when the winter's snow is still melting in patches on the ground, I'm awakened before daylight on a Saturday morning by the sound of Mr. Pickett warming up the station wagon before he takes off on one of his out-of-town revival meeting trips.

Then I hear him talking to Brother, who is asking if he can have the keys to the Cadillac because he has to run an errand. A little while later, I hear both cars leave. Without the disturbances of those noises, I'm able to fall back asleep for a few more hours.

When I get up for the day, Mizz Pickett has already left for one of her jobs as a domestic on the other side of town. After breakfast and our chores, Dwight and I watch Saturday-morning cartoons while Flo sets up camp by the record player. All day long, over and over, she plays her two new 45s, Diana Ross singing "Reach Out and Touch Somebody's Hand," and Al Green's "Let's Stay Together"—two songs I'll never be able to stand hearing after today.

Around 11 A.M., Dwight and I are on our way outside to play when the phone rings. We wait in the kitchen while Flo answers it. By Flo's answers, I can tell that it's Mizz Pickett, wanting to know if Brother is up yet.

"No, ma'am, he's still sleep," Flo says. Listening for several beats, Flo then says, "Yes, ma'am, I'll make sure he gets up and has something to eat." She hangs up and tells Dwight, "Mu-deah says go wake up Brother."

"Why I got to go?" Dwight starts to argue, turning to me as if it's my job, but I run out the door before the two of them gang up on me.

When Dwight comes outside, I ask if Brother was up yet, knowing that Mizz Pickett will hold us all responsible if we let him sleep all day. Dwight tells me that when he went upstairs, Brother said he was real tired and needed to rest more. Dwight says, "He asked me for a glass of water and I took it to him."

It occurs to me that Brother may have been drinking and maybe that's where he went with the Cadillac—to buy some of that wine he's not supposed to drink. I recall the times I've heard Brother sitting at the kitchen table telling Mizz Pickett about missing his daughter and about some of the details of what his ex-wife has done to hurt him so. Mizz Pickett likes to console Brother by saying, "It ain't gonna profit her none."

Once when she was telling him that and she caught me listening in, she shooed me away, remarking to Brother how awful I was, along with Dwight and Flo. "Ain't dese de worst niggas you ever saw?" Mizz Pickett said to him, like always.

Brother agreed that we were the worst, only later coming around to apologize: "You know how things are." That made me mad, as it usually did, but today I feel real sorry for him and wish his life wasn't disappointing him like it is. When he gets out of bed, I decide, I'll tell him that I forgive him.

At two o'clock, Flo, Dwight, and I are in the kitchen finishing up lunch and Mizz Pickett calls again. This time, she tells Flo to send Dwight up and have him bring Brother down to the phone. Mizz Pickett waits silently on the other end of the line but has to hang up before Dwight returns, unsuccessful. Nothing was getting Brother up—not Dwight turning the radio on loud, not pulling on his foot, nothing.

"Did Brother talk to you?" I asked Dwight.

"He didn't talk, but his eyes were open."

Flo gets spooky on us, saying, "Maybe he's dead."

We three inhale fast, almost gasping, then shake our

heads, laughing sort of, talking loud, "Nah, he ain't dead." Just hungover, I say to myself, following Dwight back outside as Flo starts her records up again.

At dinnertime, Mizz Pickett's daughter Teresa shows up after calling to hear that Brother still hadn't gotten out of bed. Teresa hurries in with a fishy look on her face, something I think has rubbed off on her from her husband, Tom, a carplike man with bug eyes and a thrust of his jaw that make him appear to have just eaten some questionable seafood. She grabs me by the shoulder and takes me up with her to the doorway of Brother's bedroom.

"Brother!" Teresa calls to him at full volume. "Brother, wake up!"

When she gets no response from Brother, Teresa turns and I follow her to the phone in the hallway to call her cousin Nelson. Unknown to me until now, it seems Nelson is some sort of medical authority. Nelson gives Teresa instructions, which she, in turn, gives to me. "Twonny," Teresa orders, "go put your ear next to his chest and see if his heart is beating."

Me? I give Teresa a no-way toss of my head. I like to be helpful, but I'm not about to do that. So Teresa goes back into the room and tries pushing Brother to wake him, calling his name some more. When that doesn't rouse him, she calls the police.

A handful of policemen arrive, a few black officers and one white. This is part of some progressive law enforcement policies of late. Usually the black policemen, most of them familiar, can be seen in the neighborhood in their lime-green cars. The typical white policemen are most often sent into the neighborhood supposedly undercover in unmarked cars. Of course, even little kids can spot them. Two white guys in suits in a white Plymouth pulled to the side of the road in an all-black neighborhood, just to talk to each other? And when they started mixing them up, one white undercover cop with a black one, they became even more conspicuous.

It's so ominous to have the house filling up with strangers in uniforms asking questions that Dwight, Flo, and I decide to hole up in the downstairs rest room. Without saying much, we listen to the sounds and voices outside the door, uncertain as to what's actually happening. For the last several hours, I've held on to the idea that Brother is sleeping, but the reality that he isn't going to wake is beginning to seep in. The thought of having to see them bring him down scares me. I don't want to see him dead on the stretcher, whether or not they cover him up with a sheet, because the terrible image of his form under that shroud will be just as bad.

As they came to discover, Brother had swallowed a combination of Sominex and Bayer aspirin. It was the second time in my life that anyone I knew had died. The first time was at the old house when I was four and my friend Michael's father died. They lived across the street in a yellow-and-white house where friends and family, dressed in their good clothes, gathered in what I was told was a going-away party. That's what Mizz Pickett told me, that they were going to "see Michael's daddy off."

Cars were parked up and down both sides of the street, as was a big old station-wagon-type vehicle I later knew to be a hearse. My impression at the time was that the reason for the big car was so that Michael's dad would have somewhere to put all his stuff, which he would need wherever he was going. Michael cried for weeks that his daddy wasn't coming back no more, even when I tried to say that he might. When I understood the finality of his father's departure, it bothered me for Michael's sake how everyone else's lives continued in everyday fashion.

When Brother killed himself, I had similar feelings. It seemed the world should have stopped or paused or acknowledged in a collective way that a living person was gone. But the world didn't stop. Life just went on.

Of course, at Brother's wake, there was crying and comforting and people coming in from out of town, with talk about how at least his suffering was over. In his casket, Brother did look peaceful, dressed in a nice green suit in which to be buried. His elderly mother had come to Cleveland, and I watched from my seat in a pew as she lovingly approached her son and clipped a fountain pen to his front jacket pocket. That struck me as odd—like he was going somewhere that might have some documents for him to sign. But the part that bothered me most about Brother dying was that now I wouldn't get a chance to tell him I forgave him and that I really did understand how things were.

four

In November 1973, according to my social services file, not long after I started the eighth grade when I was fourteen years of age, the state of Ohio removed Dwight from the Picketts' house. Obviously, Dwight's social workers had become aware that his placement with the Picketts wasn't working out, but the fact that I was left in the home indicates that they didn't recognize what the real problems were.

Over the past four years since our move, Mizz Pickett had once again managed to elude social services' questions about just how suited she was as a foster mother. One of her ploys was to point out to social services that thanks to her guidance, I was now doing so well (after my years of truancy and thievery), but that Dwight was so rebellious and deviant he had to be beyond help. The pattern was that just when the social services office tried to closely examine our situation in order to remove Dwight if necessary, Mizz Pickett would suddenly say that Dwight had improved and that I was the rebellious and deviant one.

In an effort to figure out the true dynamics, one of my social workers, Ms. Wood, made these observations about us when she came to take us to the clinic:

Antwone was fairly quiet during my
visit. When I mentioned his Art
work, he became a little more talkative. He brought me some drawings
to show me. His work is very
good. . . . Dwight began drawing to
show me that he was better than
Antwone. Dwight constantly insti-
gated fights with Antwone while I
was there. Antwone was rather pas-
sive. . . . Dwight was teasing
Antwone on the way to the clinic
and I could not get either of them
to talk about anything constructively.

Toward the end of her year on my case, Ms. Wood
turned her attention away from the symptoms and tried to
get to the cause, writing:

Recently there has been some question about Mrs. Pickett working. She
has not admitted this to me, but is
very seldom in the home. This should
be explored to find out what effect
if any her absence from the home have
on Antwone and the other children in
the home.

Before she could explore the question, my case was
transferred to Ms. Balestreir, whose overburdened schedule made her unable to arrange an initial visit for almost
five months. When she finally made that first visit, she
described Mizz Pickett's easygoing reaction:

Mrs. Pickett told me that it has been
her experience not to see workers
very often throughout the year. I
explained that because of my case-
load contacts would have to necessarily be limited to once every two
months or once every three months.
She said that this was perfectly
fine and that she had had no major
problems with the children.

On the contrary, there was a major problem developing
with Dwight, who, having recently met his mother, had
begun trying to run away to go live with her. Not knowing who or where my mother was or why she had never
come for me, I wasn't so sure anymore about my old
dreams of the kind, beautiful woman who would appear
at last after the awful mistake of me being taken from her
and put in the orphanage. I was older now and had come
to see that our stories weren't so simple. We had heard,
for example, that Keith's father was a white man who
raped his mother and she was unable to take care of him
as a baby. When she got on her feet enough to support
him as a single mother and Keith had gone to live with
her, he was unhappy and was left unprotected from the
predators in that environment.

Dwight's background was also complicated. He eventually knew who his parents were, but they weren't together. His father, who was different from Flo's, lived in
another state, and their mother, for whatever reason, was
unable to have him and Flo with her. In earlier years, they
used to go away overnight on the weekends to visit their
uncle Mickey.

On one of those overnight stays, I got to go to Uncle

Mickey's, too. Uncle Mickey's daughter had her own bedroom with the most ornate canopy bed, like a shrine to her. It was almost too much, I thought, but figured that was how normal people lived. Someday, I was determined to sleep under a canopy and know how it felt.

When Dwight's mother contacted social services and asked for a visit, he and Flo were both given the opportunity to visit her. Dwight was overjoyed and could talk of nothing else, but Flo said no, giving no explanations. As she was getting ready to finish high school and had plans to enter the Job Corps, I wondered if maybe it was too late to know her mother; or maybe Flo was afraid that after swallowing her anger and hurt it might come bursting out. But Dwight had never learned to swallow much, and he returned from meeting his mother as though he'd been through a religious experience. For that one afternoon, it seemed that he had been purged of his rage and he looked at me with a soft, dreamy smile, saying that he loved his mother. "And," Dwight said, getting down to what really mattered, "she's pretty."

Even though Dwight's mother lived on Hough, not the best neighborhood, some distance from the Glenville area, Dwight began walking over there to see her. Sometimes she wasn't there or couldn't let him in, but he would hang out on the street, waiting for a glimpse of her. The more he wanted to be with her, the more he couldn't, the angrier he became at the Picketts' house, the more the Picketts took it out on him physically and psychologically.

After Brother died, the Pickett twins usually took turns coming to stay with us when Mr. and Mizz Pickett were away. When it was Mercy's turn, Dwight got into a physical fight with her before I returned home from being out one Saturday. For the life of me, I couldn't figure out why Dwight chose Mercy to fight—the nicest one of all. If it had been Lizzie, I would have been cheering him on. But for him to hit Mercy hurt me. What it was about, I never knew, only that when I arrived home, Dwight had disappeared. As had become our routine, I went in the car with Mercy and Teresa and drove over to Hough to look for Dwight. In the past, he'd been known to be over there, but that day, after hours of searching, we found neither hide nor hair of him.

As the days went by, I became preoccupied with the worry that Dwight wasn't coming back at all. As much as he annoyed me, his absence as my comrade in arms in the Pickett household left me stranded on the battlefield.

My caseworker, Ms. Balestreir, who had described her three previous visits as uneventful, made these observations at the time:

> My most recent visit centered around a request by Mrs. Pickett that I speak to Flo and to Antwone about their brother, Dwight, who had become AWOL as a result of a fight with one of Mrs. Pickett's natural daughters. I talked to Antwone and was pleasantly surprised by this child's amount of warmth and understanding. The disappearance of his foster brother had obviously gave an impression and an effect on Antwone and he was quite willing to discuss his relationship with his foster brother and their constant bickering and fighting.

While Ms. Balestreir was one of my first social workers to see beyond my shy, more watchful side, any opportunity to follow up with me was lost when five days later my case was again transferred. By that point, I was terrified about Dwight's disappearance. Unable to sleep at night, I sat up worrying about him, listening to the ghostly sounds of the house at night. Sometimes I imagined hearing footsteps coming up the basement steps into the kitchen; once or twice I thought that I even smelled toast and butter and that I heard the refrigerator being opened and closed.

In the meantime, I started getting into trouble for missing loaves of bread and dirty lip prints on the milk carton.

"You low-down nigga," Mizz Pickett disdainfully greeted me one afternoon. In her hands she was holding a watermelon half—Dada's watermelon—that had been scooped in two pulpy places by a human hand about the size of mine. She demanded to know, "Do I have to put a lock on the 'fridgerator and pass out keys?" Before I could defend myself, her hand flew up, "Ah-ah-ah, don't even start ya lie." And she waved the watermelon in my face, taunting, "I would make you eat the whole thing but you might just enjoy it!"

After two weeks, neither the Picketts nor the social services people were able to find Dwight. Then one evening Mizz Pickett was down in the basement doing laundry with Laddie sniffing around at her side. A German shepherd, and a female dog as well, she was ineptly named Laddie—so I thought—by Mizz Pickett, whose dog she was. As it was later recounted, Mizz Pickett was standing at the washing machine when Laddie started barking like crazy for no apparent reason. Going to investigate, Mizz Pickett went over and pulled back a mound of junk—old rugs and boxes—and there he was, sitting there, smelling and looking dirty and thinner, like Oliver Twist, my long-lost foster brother.

Dwight had been living down there the whole time, slipping out during the day and keeping himself out of sight, then stealing back inside at night to sleep. He couldn't really bathe or change his clothes, and all he ate was what little he could forage quickly from the fridge—milk, toast, and watermelon.

The episode clued me in to just how much pain he was in. He actually preferred living like that to what had become, for him, the torture of life with the Picketts. In many ways, Dwight understood much more than I how poisonous our situation was; and, long before me, he got the whole idea that there could be another home for us that was more legitimately ours—a place that existed somewhere else outside the Pickett realm. Dwight was the first to understand and rage against the way that Mizz Pickett cared for us; he knew it wasn't a natural coming about of love for a child because of our humanity or who we were, but because of the habit of us being there, the way she loved Laddie for being there all those years.

This sentiment was echoed throughout my file as case-workers described Mizz Pickett's reason for wanting me in her house:

```
She said that at times Antwone acts
up, but since she has had him so
long, she'll never let him go.
```

Dwight was reacting in a normal way to all those years Mizz Pickett had spent telling us that the only reason she kept us was to pay her notes. He must have recognized that every time the social services check arrived, it reminded Mizz Pickett of how cheap we were as human beings. Dwight got all that, and I didn't. At the same time, I coped better by escaping into my imagination and living there, buoyed up by imagined love. Dwight couldn't subsist on false love; he needed the real thing—or, as he might have put it, the real shit.

That was his phrase the time he came home with some weed wrapped up in tinfoil. I'd never seen marijuana before; he was calling it herb. "The real shit," Dwight said proudly, "Panama Gold. And guess what else? I'm gonna *smoke* it."

Sure enough, he left the house and came back some time later, looking red-eyed but relaxed. Since he was somebody who liked to explore different turf, Dwight's starting to get high didn't shock me much. Maybe it gave him a temporary release, like some of the other stuff he liked to explore—books and a range of music, mostly white rock 'n' roll, that a lot of people in our all-black neighborhood thought was strange. Dwight didn't care. He wasn't ashamed to be into the James Gang and Deep Purple or to put a radio in the house window and have AC/DC blaring out.

Mr. Pickett came home to just that once, with neighbors out in their yards looking irritated, but he didn't say a word. Maybe he understood Dwight's need to be a teenager; maybe he even remembered his own adolescence. But by ignoring the incident, what Reverend Pickett failed to do was give Dwight the attention he was craving from a father figure, a father figure who didn't give anybody much attention. Mr. Pickett didn't mean anything personal by it; it was just the way he was.

Some of Dwight's music wasn't so bad, I thought. I liked Jimi Hendrix a lot. But getting high wasn't for me, for a few reasons, not the least of which was the fear of Mizz Pickett's wrath if she caught me. Typically, of everyone she could have accused of smoking reefer, she picked me. That was ironic, since I was the only teen or young adult in the house—including some of her own offspring—who wasn't getting high now and then. She may have suspected that Dwight was the one smoking dope, but was too busy riding him about other aspects of his alleged deviant behavior to attack him for that as well.

Mizz Pickett couldn't see that, aside from wanting love, Dwight just wanted to be a regular teenager—to stay out later at night with his friends, go on dates with his girlfriend, and, as Mr. Pickett did understand, rebel through personal tastes in music and reading. Dwight had a girlfriend for a short while, but it could never be full-fledged because he could never interact with her normally—not with Mizz Pickett in the mix. With a girlfriend, there would inevitably be a scene that involved Mizz Pickett finding out or, worse, the girl's parents wanting to meet our parents: a mess, a confrontation with Mizz Pickett calling their daughter a heifer, a page of embarrassment in the already humiliating chapter of our misbegotten teenage years.

So Dwight and I took it out on each other, with teasing and jousting. He had only to spot some pretty girl and say, "Hey, man, that your girlfriend?" to make me mad enough to punch him. But because Dwight fought to hurt—grabbing my mouth, gouging my eyes, twisting my arm—I preferred to give him my angry look and leave it at that. He'd drain my angry slit-eyed glare of its power, saying, "Don't do your eyes like that. It makes you look like a girl."

For fun, Dwight and I played the Dozens with each other. He was usually better. There was one day, however, when he was reading statistics from a WWF magazine and asked me, "Guess who got the biggest chest in professional wrestling?"

I smiled and said, "Yo' momma."

As soon as we started laughing uproariously, Mizz Pickett stepped into the doorway, leaning into the room as she sang out, "Lawwwd, have mercy. These niggas in here playin' the Dozens." She leaned down even lower, her nose scrunched up in disdain. "Don't you know, that's the dirtiest game you can play?"

The dirtiest game? She had been playing the Dozens with us for as long as I could remember. I thought of some other unsavory games—tying kids up and leaving them in the dark for hours, beating a child until he passed out, evading questions from social workers and the government agency responsible for that child.

I thought of how Mizz Pickett had praised me for taking the initiative to earn my own spending money by doing yard work for neighbors. After using my first bit of savings to buy myself a shirt, she couldn't say enough about how resourceful I was. Then she started hiring me out to some neighbors and having them pay her directly, money I didn't receive.

Given my survival skill of silence, having learned to play the innocent to the Pickett audience, it wasn't my style to challenge Mizz Pickett's opinion of just how low the Dozens could be. But after Dwight started to leave by running away, he no longer held back from confronting Mizz Pickett about things like that. Unlike me, Dwight was unable to accept the status quo and make the best of it.

My way of coping was reflected in my file, even though it was shaped by Mizz Pickett's input, when yet another caseworker, Ms. Edwards, was led to this conclusion:

Antwone seems to be very dependent
on Mrs. Pickett. He makes an effort
to please her and she rewards him for
this.

Apparently, since the subject of Dwight and his mother
had become such a hot issue, Ms. Edwards raised it with
me, writing:

Antwone has not seen his mother
since he was five years old. When
caseworker questioned him, he said
he did not remember ever seeing her.
He was quite uneasy during this conversation. Mrs. Pickett told caseworker Antwone was quite upset when
he returned home subsequent to interview. She said he woke up in the
night crying and sobbing. He would
not quiet down until she promised he
would not have to see his mother.

Mizz Pickett had made up most of this. The truth was
that I was ambivalent about any mention of my mother.
But for all my years of nightmares and waking up in
tears, Mizz Pickett never comforted me. Several more entries follow that describe my adamant stance about not
wanting to have anything to do with my mother; each
entry was based on social workers' conversations with
Mizz Pickett. No one ever told me that during this period,
my mother had, in fact, been in touch with social services
in the hopes of meeting me. If anyone had, I may well
have concluded that I wasn't ready to meet her; on the
other hand, I was never given that choice.

It's possible that Mizz Pickett was using her bizarre brand of psychology to prevent me from becoming interested in my mother because of the trouble that Dwight was giving her. Not only was he confronting her over various concerns, he had also zoomed in on an argument that hardly made any sense but was based on the fact that the Picketts were getting up in years.

"Mu-deah," Dwight said the first time he broached the subject of going to live with his mother, pelting her with questions, "you're not gonna live forever. When you go, where am I gonna be? Am I gonna live with your kids? Am I gonna be a part of this family?"

"Sho' they love you, jes' like I do. Ya been wit' us all dese years, ain't-choo?" Mizz Pickett asserted. But she had made it all too clear over the years that we weren't her children, so how could we feel a part of her family?

Dwight continued, "You ain't gonna leave me nothin' in your will."

"Well, you might jes' get somethin'."

He went down a list of household belongings he doubted he'd get, including a rickety old boat they probably couldn't pay to give away. I couldn't believe Dwight was saying all this. In my mind, I couldn't care less if I got anything from them. No doubt he felt the same way, but this was his attempt at logic for wanting to go be with his mother. He couldn't simply say what the real truth was: "I love my mother and I need my mother, my own mother." If he had, Mizz Pickett would've said, "Nigga, wha'choo know about love? That woman don't wont-choo. That's why you standin' here in my house."

After Dwight brought this up so many times, that was exactly what Mizz Pickett did end up saying to him. For all the meanness she had used when saying things like that in the past, at this point her tone was actually thoughtful, as if she was trying to be sensitive, in her own way, and have him face the uncomfortable facts.

Away from the Pickett house, I had adjusted to junior high school and was making new friends. At home, Dwight's drama continued to play out with more disappearances, more scenes, more anger and desperation. If ever a person needed one thing from this world it was Dwight, and his need was love. There were many things I longed for that Dwight could have done without. Not love. He was like a huge, intricate human mechanism with its one missing gear that kept it from working. With love, Dwight could have been someone to impact on the whole society. He was that special and that smart, like any one of those enlightened individuals who come from the most meager of conditions. But without love, Dwight didn't work, and I believe everyone lost out.

Finally, because his mother couldn't take him in, it was arranged for Dwight to go live at Boys Town, a group home for troubled or hard-to-place adolescent males. Though it was on the edge of the Glenville area, we never saw him. I would later hear he'd been running away from Boys Town until, at last, he was allowed to stay with his mother. When that didn't go well, he was sent to live with his father. No longer a ward of the state, he wouldn't stay there long either before being unleashed, unprepared, into the wilds of the world at large. From there, I'd lose Dwight's trail for many years, eventually catching up with him as stories filtered through about his crime spree that would cause him to spend most of his life in prison.

"Such a shame about Dwight," Mizz Pickett would

later remark, "the way he jes' threw his life away," never realizing that it was she who threw his life away.

Flo's fate would take her in a different direction. After the Job Corps, she never married and never had any kids. While I don't know how she experienced what we went through, I wouldn't doubt that her later choices had to do with the earlier rejection.

Every person has his or her own level of pain tolerance. You can put a little weight on someone and crush him; you can put a lot of weight on someone else and not crush him. Maybe I learned to be like an orange, something that you squeeze and squeeze only to separate the juice from the pulp, without taking away its basic substance. I knew how to do that; Flo may not have; Dwight definitely didn't.

Two weeks before Thanksgiving, with wet cold winds streaming off the lake, Dwight left the Pickett house at night after dinner. The last I saw of him, he and Mizz Pickett were standing in the hallway upstairs, under the dim light of a wall fixture, next to an old television console with wooden doors that opened and shut, selecting clothes for Dwight to take with him. Mizz Pickett was acting warm and courteous as she folded and piled the clothes on top of the console. Since Dwight barely had any of his own clothes, several items of my clothing were stacked in that pile. Too many other emotions were darting through my senses for me to care much about that.

Then Mizz Pickett took the stack of clothes downstairs, singing sweetly up the stairs, "Dwi-iight, it's time to go now." We heard a car's engine revving in the driveway— probably Mercy or Teresa, or another one of Mizz Pickett's kids, who would be driving Dwight over to Boys Town.

Dwight turned to go, a look of such victory on his face—as if Abraham Lincoln himself had set him free. Then he turned back to me. He was fifteen, still wiry, good-looking, shorter than me by now as I, age fourteen, pushed my way to my eventual height of six feet.

"See ya, man," he said, and took one last look around.

After I watched him fly down those steps, I turned to walk down the hall to Brother's old room, where there was a window looking out onto the street. My steps were slow as a new heaviness came over me. From the window, I saw the car pull out of the driveway and take off down the street, fading into the darkness, and Dwight disappeared from my life.

Nigga!" Mizz Pickett's holler pierces through the fragile dome of my sleep early this spring Sunday morning. She has a new plot percolating to fill the beds vacated by Dwight, Flo, and Brother, a plan for the county to send her money for boarding mentally handicapped individuals. With that in the works, I'm the only boarder she's got, and she's determined to get her worth out of me.

At the top of the stairs, I look down and see her on the landing in her springtime church attire: light blue pleated skirt, white blouse and matching cardigan, heavy black old-lady shoes, and her trusty black patent-leather purse slung on her wrist with her palm turned up. Though Mizz Pickett hasn't stopped berating me for my refusal to go to church, she has stopped locking me out every Sunday. Instead, she leaves me with a list of chores each week as she does this day, removing her glasses and cleaning them with her own saliva while she talks. "Oh," she adds, like she almost forgot, "when you finish, don't run off." Putting her glasses back on, she squints at me ominously, telling me, "Tom's stopping by with somethin' real important, and you got to be here to put it up when he drops it off."

After she and Mr. Pickett leave, I obediently tackle my chores first—working quickly and thoroughly, as I have learned to do—and then settle in for some TV watching. *Batman* is on when I hear Tom's car pull into the driveway. The fish-eyed Tom, now Mizz Pickett's ex-son-in-law, Teresa's ex-husband, leers at me when I open the kitchen door, saying nothing as he pushes a heavy, lidded roasting pan into my arms.

"Thanks," I say, not knowing what I'm thanking him for or caring much to find out. Tom leaves. I set the roasting pan on the table, shut the door, and go back to catch the end of the show.

Later in the day, on my way out to join up with some friends down the street, I decide to check out the contents of the delivery. Gingerly I pick up the lid and there, grinning right at me, is a disembodied pig's head! In horror, I slam down the lid, sending the roasting pan off the table and onto the floor, spilling pig-head juices everywhere. Fortunately, the head itself doesn't fall out, and I'm able to wipe up the mess without leaving any trace evidence.

Upon my return in the evening, Mr. and Mizz Pickett, still in their church clothes, are sitting at their Sunday feast of neck bones, collard greens, rice with gravy, black-eyed peas, and, of course, the Reverend's ever-present great German beer mug, studded with moisture as though sweating, brimming with Kool-Aid.

First thing out of my mouth is the announcement, "Tom brought that roasting pot over with a pig head in it!" I'm sure they're as shocked as I was.

Mizz Pickett only nods with a thrill, saying, "Yeah, I'm gonna make us some hog head cheese."

"Hog head cheese? What's that?"

"You done had some."

"I have?"

"Sho' you have, and you love it."

My stomach begins to churn, remembering the time she served chitterlings, pig intestines, and I saw that they were still working on the plate, performing their previous natural digestive duties. Hog head cheese, Mizz Pickett explains, is when she grinds it all up into a sausage with red peppers and spices and makes it into a gel, like Spam, that you slice up into pieces marbled with cartilage and make into sandwiches. As she tells me this, Mr. Pickett continues to chomp and chew on his dinner, the way he always eats—like an eating machine.

Mizz Pickett accompanies him with her own eating noises, her dentures knocking together as she chews, making the same hollow sound as ladies' heels echoing down an empty corridor.

That sound puts me over the edge and in this moment I make a firm decision. I deliver the second major announcement of my life, probably more words than I have ever spoken all at once to the Picketts. My gaze steady, my posture erect, my voice direct, I say, "I'm not eatin' no hog head cheese."

They both looked puzzled. And I'm not done. "No raccoons, neither, no more deers and no possums." They both jump back in their seats, visibly shocked hearing this from me, the painfully shy Twonny. You-all heard me right, I think but don't say. Then I lean in with my summation: "No more pig ears, pig feet, pig tails, chit'lins, and no turkey necks." When I finish, I make another decision—that I hate grits and ain't gonna eat them no more, neither.

Perplexed, Mr. Pickett keeps eating but speaks between mouthfuls, "Say ya ain't?"

"No, sir," I say.

Mizz Pickett, really baffled, lets the fork slip from her

hand so it clangs against the plate, and she stares for several seconds before saying in the kindest tone I've ever heard her speak, "Well, what-choo gone eat?"

Mr. Pickett, now sucking on a neck bone, looks up at me and back at her and says, "That boy don't know good eatin'."

Mizz Pickett picks up her fork and scoops a pile of slippery collard greens as she concludes, "Well. That don't make me no mind, that jus' make mo' fuh us."

And I think to myself—*good!*

Before I went to Franklin Delano Roosevelt Junior High School, which was also on Parkwood, a mile or so down from Parkwood Elementary, I worried about the rough kids who went there. After only a few days, those fears vanished as I realized many of my old friends were making the transition with me—Michael Shields, William Howell, Fatima Doyle, Donald Shumpert, plus lots of familiar faces from the neighborhood. These would eventually include Fat Kenny, so nicknamed for his hefty size, and his sister Sonya, both casual, easy-going kids.

On the other end of the personality spectrum was my friend Jessie. Bold, unpredictable, and fearless, he was the type of friend that was good to have because you wouldn't want him as your enemy. Though he was different from Dwight, he helped fill the void created by my foster brother's departure. Jessie's intense looks reminded me of a miniature James Brown—a medium boxy afro, with a prominent brow above his brooding eyes—as solid and menacing as a grown bull. Never wise enough to know when to be scared, he was always doing crazy, reckless things. Over time, I saw how dangerous that lack of fear was, and it taught me the value of being scared.

He would take on two guys both bigger than him who were getting ready to jump him, like Jesse James—and always win on bravado alone, enjoying every moment of it. He'd fill up on half a loaf of bread before going up to 105, give a wino money to buy him a bottle of Night Train, and then go sit on a fire escape at Parkwood Elementary, drinking his wine or smoking reefer, or both, in the middle of the day. Other kids did that, too, but Jessie did it a lot. Sometimes when I passed by, I'd see him up there all alone—high, which became his preferred state of mind, and dancing to the music in his mind.

When I wasn't in school or hanging out with Fat Kenny, Jessie, Michael Shields, or some of the other guys, they all knew where to find me—outside in every kind of weather, walking up and down the street, mainly in the vicinity of the 900 block of Parkwood. Michael used to say, "There goes Fish again, out patrolling the neighborhood." But he never knew why I did it; he never knew the full extent of my passions.

My patrol was mapped to allow me to walk past the houses of various girls who had captured my attention—girls who actually knew my name, a few who liked me in return but who would never have been so forward as to tell me so. Not that they were afraid to strike up a conversation with me; but they all knew, as I did, that given my inordinate degree of bashfulness, just the words, "Hello, Antwone," or "Hi, Fish," would easily cause me such bitter discomfort and shame that I'd be forced to cancel all patrols till I recuperated from the effects of such warm hellos.

Then a new girl arrived in the neighborhood. There was a flutter mixed with a panic the first time I laid eyes on her, and her face lingered in my mind constantly as I invented every perfect scenario starring the two of us. Early on, I got that this wasn't like before with the girl in the yellow dress where it all existed in my imagination; this time, it was having a physical effect on me.

Five, six, ten times a day, I walked past the house where she lived. In four feet of snow, in the driving rain, even through the most fierce midwestern tornado. All for the chance to get a glimpse of her. Or, hope against hope, that she'd look out her window and notice me and I'd be planted in her memory forever. In art class, where I was a superstar and practically alone with her, it took everything within me to pretend that being that close to her didn't matter at all; all the while, my mind was doing somersaults and she was all I could think of. Try as I might, I couldn't take my eyes off her.

Her name was Freda Smolley. With her sister, Mona, their two brothers, and her mother, Ann, she had recently moved into a two-family home on Parkwood, a few blocks from the Picketts' house. By neighborhood standards, Freda was pretty, but not as pretty as Mona. By my standards, Freda was more than pretty. Petite and coy, yet tough (not as tough-acting or flirtatious as Mona), Freda was a little darker than a light shade of brown and wore her hair rolled back in the shape of an afro. And she had a slight overbite, a small imperfection I found absolutely attractive. Over the years, she changed as she grew up, but my image of her was always as I first saw her at eleven years old, a real girl next door in faded blue jeans, a clean white T-shirt, and black desert boots. That was how we all dressed, the uniform of the day.

Ron Banks, Larry Demps, Willie Ford, Lenny Mayes, and L. J. Reynolds comprised the Detroit singing group known as the Dramatics, and they sang an anthem for every round of young love I ever had. "Whatcha See Is Whatcha Get," "Fall in Love, Lady Love," "Fell for You," "Be My Girl," and "Shake It Well," to name a few. Whatever information I had about romantic relationships, I got from the Dramatics, feelings and ideals that drove me mad with love for Freda. No passing flame, no fleeting fantasy, she dominated my dreams for years to come, even long after we were separated.

For all the unrealistic expectations of love that my romanticism caused me, those musical sensibilities began to pay off when the school gave me a job playing records during the lunch-hour recreation period that all students could attend in the girls' gym. This made me something of a big man on campus, sort of a celebrity, and I soon acquired a great record collection. Sometimes the school gave me money to buy the records; most often I spent my own earnings on them, earnings I made sure Mizz Pickett never found out about.

Before long, Freda was asking to borrow records—Marvin Gaye, the Dramatics, many of my favorites. Then she asked to borrow a brand-new arrival, "Do It Baby" by the Miracles. Normally, I wouldn't let anyone take the records home, and especially not that one, because it belonged to the school. But I wasn't going to pass up an opportunity to please Freda. I panicked when she didn't bring it back the next day and all the kids were requesting it. Luckily, I placated them with "Jungle Boogie." The following day Freda returned the Miracles' record to me, saying so cutely when she handed it back, "I love this song, I had to keep it two nights." And then she added, "Thank you, Antwone." Hearing her say my name, instead of my nickname, felt real good.

I shrugged and nodded, as if it was no big deal. But it was to me, and this was about the closest I came to acknowledging how I felt about her. Although there was the time I mixed her paints in art class. It used to infuriate Jessie, how Freda and Fatima would both ask me to mix their colors. He was even madder the day that Freda's sister, Mona, asked me to draw her; apparently Mona was interested in me and not him—just like Freda.

As mad as he could get, though, it was a credit to Jessie's restraint that he never teased me about the extent of my terrible shyness. A worldly kid, to say the least, by the time he finished junior high, he had already fathered a child. When I found out, I was flabbergasted. "Golly, man, you're having s-s-sex?" I asked, having trouble just saying the word.

Jessie tried at times to cultivate my worldliness. Once when we went to a basement rent party at the Smolleys', he attempted to get me drunk, assuring me it would help with my inhibitions. I got buzzed on a capful of Night Train that burned my throat and made me feel silly and light-headed. Tripping down the street next to him on our way to the party, a loaf of bread under my arm, a brown bag with the wine in it under his, I waved my arms trying to get my balance, saying, "Everything's spinning!"

"Eat some more bread" was his matter-of-fact response. But the buzz didn't make me bold. Not even when we got to the party and Freda was at the door collecting the fifty cents admission from each person and when it was my turn, grabbed my hand before I put the money in the can, telling me I didn't have to pay. Buzzed or not, expressing my feelings in return was simply impossible.

But somehow Freda understood that mine was not your run-of-the-mill shyness, that it was on the level of a brutal terror in my bones; and she never pushed me beyond my comfort zone. As normal as I tried to be, I couldn't undo the fact that most of the human touch I'd experienced in my life had hurt. So anything physical with Freda, even holding hands, was out of the question. But there were times when we did walk together, just the two of us.

Going to school in the mornings, whenever I approached Freda's house, I'd slow down and sometimes see her peek out the window, spot me, and moments later appear on the street. We talked along the way, but never about anything meaningful. And yet, I couldn't imagine two people more compatible. We were meant for each other, clearly, and I saw us married one day in the not-too-distant future, living in our own house on Pasadena, a block away from the Picketts, where the street was lined with weeping willows—my favorite kind of tree. It didn't occur to me to imagine us farther away, not because I felt in any way connected to the Picketts, but because the world within the confines of the Glenville area of Cleveland was the only one I knew.

As my friends became aware of how I felt about Freda, without embarrassing me, they cheered me on as their most humble hero. But with my few items of old clothes, it became harder and harder to look heroic. Then I discovered spray starch. Every morning, I got up early and starched my clothes. It made a huge difference. My reputation grew as a person of impeccable appearance. Given that, on top of my job playing records, I was enjoying real popularity.

Besides being DJ during rec, I also became the projectionist for special monthly movies in the auditorium. The day we were showing *The Learning Tree,* as was my routine, I stood up in the balcony and made sure the projector was properly pointed at the big screen down on the stage. Looking around the auditorium filled with noisy, overly energetic students, I saw my group down near the front, where Sonya, Jessie, Mona, and Freda were saving me a seat. Among them was a particularly pretty girl whose looks rivaled Mona's, and after I turned on the first reel and hurried down to join them, this pretty girl announces she has to go to the rest room.

Pushing her way down the aisle, she plops right into my lap as if by accident. The guys are eyeing me enviously; the girls are annoyed. Trying to keep from blushing, I'm feeling *gooood*—like, hey, I'm *the man*. She bounces back up, giggling, and moves on, returning quickly after that, not trying her trick again when she crosses by back to her seat.

By now, we've all settled into watching the movie, when suddenly we hear someone call in a singsong taunt from the back of the auditorium, "Norma Jean!"

At hearing the name of her mother, Sonya shouts back, "I ain't playin'!" Next thing we know, this shorthand version of the Dozens ripples through the place as mothers' names are being called out right and left. When somebody says, "Ann!" Mona spins around, saying, "Leave my momma out of this!" and she and Freda jump up and start cursing.

Hilarious laughter rocks the room, and we're all falling out of our seats, the music of our carefree adolescent humor pouring into the air and condensing into clouds of amusement that rain down on us in the dark in front of the movie screen. Thankfully nobody tries to get me by calling "Miss Fisher!"; and I'm really glad that no one knows to call out "Tha' Lady Isabella!" For a second, the thought of Mizz Pickett plays a sour note in this symphony of good feeling, reminding me that I'm different from everybody else, that I have no mother of my own, no mother whose virtue to defend. But I shut off such thoughts, grateful for the fellowship of the afternoon, savoring the knowledge that my friends genuinely like me.

A week later, I have another occasion for gratitude. This day, a tough kid has promised to meet me after school and fight me. When we get outside, a large throng of fellow students has shown up to watch. So the guy makes the first move, grabbing me forcefully and shoving me back. I stumble, regain composure, and move toward him to push back when I look up and see he has disappeared. The crowd just grabbed the guy and sucked him in. The next day, with two swollen eyes, covered in bruises, the kid approaches me, saying, "Let's be friends." Apparently, my public didn't want anybody messing with Fish.

More than ever, school was my haven away from the Picketts. It was my friends, our camaraderie, our activities; it was art, learning new subject matter, a new sense of academic accomplishment, and adjustment. But mainly because of worsening conditions at home, I couldn't ever be truly happy—not in any way that lasted longer than the school day itself.

In my community, I felt unhappiness all around me. In the shadow of Vietnam, the race riots of the earlier decade, the growing spread of drugs and guns, and the overwhelming specter of poverty, the Glenville area was showing the wear and tear being taken on black ghettos all around the country. I could see it in the lost eyes of men and women standing around on 105, a street I'd practically learned to walk on, that I now learned to avoid. I could even see it with things that were happening to kids I knew from school.

One of them I didn't actually know well. But I wanted to know him because, for some reason, all the girls were crazy about him. What his secret was, I was dying to know. Then one day, lo and behold, we end up walking home from school together and we see this cute girl walking toward us, holding her schoolbooks in front of her, and he turns to me and says, "Watch me whoop my game on her."

This is an answered prayer and I watch, standing back a little, as he goes up to her. She looks at him sideways, like she's not interested; he looks at her, full front, real interested. And I'm holding myself back from jumping up and down, knowing in one second I'll be given the key to all my future girl endeavors. Then he says: "Hi." Cool, smooth, simple. Just "hi." Before she can even answer, he gestures to me to follow him. A few paces down the street, he gives me a smug, proud look, sure that I'm impressed. And I am. Just "hi." It's a whole new world of magic.

When we get to my block, I say good-bye before we get to the Pickett house, not wanting him to pass by with me—just in case the new boarder is having a mental attack on the front yard. "See ya, Fish," he calls, and continues on to his house a few streets over.

The next day, I look for my new friend in the halls but don't find him. Not this day, nor the next. After asking around, I find out that he was shot and killed two hours after he and I walked home together, the same day and the last day I saw him whoop his game on the cute girl. A numbness sets in. I can't believe he could be so alive and present one day, then gone the next. For some time, I go by his locker on a daily basis until I can intellectually accept that he isn't coming back. When I do, it bothers me, as it did with Brother dying, that the world doesn't stop and mourn him, that school and life go on, as if nothing has changed. But to me it has.

When I asked around, I heard that the boy had been

shot in his own backyard by an uncle whose mind had been messed up in Vietnam; shot in the chest as he played, maybe dreaming of a future he'd never have.

A few months later, Donald Shumpert, an impressionable kind of kid I'd known since Mrs. Profit's class, shot and killed himself by accident. Donald always seemed to me to love the idea of his older brothers—their dangerous, glamorous world—and maybe wanted to be like them. We'd seen him try to be cool by bringing marijuana and cocaine to school, but then not really knowing anything about the drugs. One day after school, according to what we heard, Donald found a gun lying around at home, thought it was unloaded, put it to his head, and pulled the trigger.

When Donald Shumpert died, I again expected some sort of pause in the daily hum, a time of silence, at least, in which to notice that something terrible and senseless had happened. But there was no interruption, no change. The sun rose and set the same, evening followed afternoon, morning came before noon, then night. Life went on.

My social worker Jill Edwards stayed on my case longer than most of her predecessors. With shoulder-length straight blond hair, nice clothing, and a warm, professional manner, Ms. Edwards was young, as all my caseworkers tended to be, and, at first, after Dwight left, she seemed to be swayed by some of Mizz Pickett's old tactics designed to minimize her visits. Jill Edwards initially reported:

```
Antwone has made a total adjustment
to this foster home. He wishes to re-
```

```
main with the Picketts and they have
told him he will remain in this house
as long as he wishes.
```

My social worker stated that, according to her primary
source, Mizz Pickett, my earlier problems had all been
caused by Dwight and that I was much more relaxed and
secure since his removal. In fact,

```
Mrs. Pickett has told worker she
would like to adopt Antwone but
should we not be able to get perma-
nent custody, would want Antwone on
a long term foster care basis. . . .
He excels in art and Mrs. Pickett is
very encouraging. She wants to send
him to Art school when he is older
as she feels he has real talent. As
the Picketts are in a financial po-
sition to do this, it is hoped
Antwone will take advantage of this
opportunity.
```

When Mizz Pickett mentioned the part about helping
me go to art school, I believed her; for the moment, she
probably meant it. So when she brought up the possibility of adopting me, I warily went along with the idea. I
say warily because I wasn't sure what it meant. Adoption
was one of those legal terms no one ever explained to me,
just as no one ever explained what it meant to be a ward
of the state. Besides that, a subsequent exchange with
Mizz Pickett made me question her sincerity about wanting me to be a bona fide member of the family.

This is a Saturday morning that has begun with the

pleasing smell of pancakes on the stove and the sound of Mizz Pickett singing. Knowing her goodwill can change as radically as the weather, I guardedly come down to eat, greeting her with "Good morning," as I'm required to do every day. After enjoying my breakfast and performing my household and yard duties with even greater precision and speed than usual, I can see Mizz Pickett is impressed. She tells me so and then says it happens that one of her nephews is coming by and he'll be impressed, too. Thinking this is an opening for conversation, I show my interest and curiosity by asking which side of the family that nephew is from.

In a split second, Mizz Pickett loses all her chumminess, barking, "None of your business! And why you wanna know?" Then she dismisses me with a scornful jerk of her head and purse of her mouth.

Shortly thereafter, when the Picketts are besieged by another tragedy—the death of their son Junior—I am not included in the wake or the burial. Besides Mercy, Junior was one of the only other Pickett kids to show me kindness. He thought enough of me, in fact, to hire me to look after an apartment building he owned. A truck driver by profession, Junior was often away on long hauls, leaving his wife and child alone at night. From time to time, he even asked me to stay over and help look out for their well-being. On one of his out-of-town hauls, Junior was asleep in the sleeping compartment while another trucker drove. It was the middle of the night and the man driving fell asleep behind the wheel, causing the big rig to jack-knife and crash off the side of the road. The driver survived, but Junior was killed instantly in his sleep.

Since I feel there has been a personal connection between me and Junior, it hurts to be excluded from his funeral. It hurts, too, to see Mr. Pickett struggling hard to maintain his usual calm, even though I know the loss of his son, his namesake, is as devastating to him as it is to Mizz Pickett, who cries for days. I have the wish to comfort him, to let him know that while I'm not his son I'm real sorry about Junior. But, of course, it's not my place.

Still, I worry about Mr. Pickett, the one sane person in the household of crazy people—literally crazy, now that the boarders have taken over the upstairs—whom I have come to see as a sort of ally. What I base this on is somewhat sketchy but it could be that I admire him. I admire that he's almost seventy years old, still as strong as an elephant, working day in and day out. I admire that he doesn't sleep in the same bedroom as Mizz Pickett anymore, having turned the den downstairs into his room, which he keeps padlocked when he goes out.

Maybe what I most admire is his calm, like the time when I was a teenager, walking barefoot in the backyard, and stepped onto a six-inch rusty nail protruding from a board. The nail went all the way into the sole of my foot, a pain that sent shock waves of blazing hurt up my leg and into my brain. Hearing my uninhibited yell, Mr. Pickett looked up from his toolbox and ambled calmly in my direction, instructing me to lean face forward against the house with my hands outstretched. With a firm grip, he yanked the board and nail right out of my foot, seemingly yanking a flow of tears from my eyes that I couldn't restrain. For a minute, Dada stood there studying the bloody, rusty nail. I waited for him to say something, but he didn't. The problem, in his mind, apparently, was solved. Later, I did wonder if a tetanus shot might not be a bad idea, but by then I was well past the stage of developing any flesh-eroding disease and concluded he was all the doctor I needed.

To me, the incident showed not only his calm, but also

made me feel we had a special bond, having survived crisis together. So when the issue of adoption is raised, I'm not unhappy with the prospect of becoming his adopted son.

It may have been at this point that Ms. Edwards noted in my file:

```
As Antwone and the Picketts both
want to begin adoption proceedings
and natural mother has no interests
or plans to care for Antwone, CW will
petition Juv. Ct. for permanent cus-
tody.
```

Then an interesting exchange between me and my potential adoptive father takes place. It's late afternoon, after school, in early spring, still cool out but warm enough for grass to be growing again and lawns to need mowing. On my way out of the side door and down the driveway, off to patrol Parkwood, maybe to run into Freda, as if unplanned, or maybe to find Jessie and Fat Kenny and Michael Shields, I hear Mr. Pickett's voice ring out, "Boy! Boy!"

I turn around and see him at the top of the driveway stooped over one of his many lawn mowers. His thick body is swaddled in grass- and mud-streaked khakis. The late-day sun reflects the sweat that's dripping down his bald shiny head onto the many rolls of fat on the back of his neck—some call his a deacon's neck—which looks to me like a pack of hot dogs.

Pointing to myself, I call back, "Are you talking to me?"

"Yeah, you a boy, ain't-cha?" he says, without mirth or malice. "Come on over here and help me change de oil in this-here mower."

"Yes, sir." I walk up to him where he's stooped and give him a hand. Once the oil change is complete, I ask if he needs me for anything else and when he says no, I start walking back down the driveway.

"Boy!" his voice calls out to me again.

I turn back to him. "Sir?"

He pauses a moment and looks at me with a serious expression as he asks, "What is yo' name?"

Confused, I say, "My name is Antwone."

"Oh." He pauses again. "That's a nice name."

Here I'd been living with him all these years and I thought we were getting along so well, with our special unspoken bond, and the whole time he didn't even know my name.

It was only days later when Mizz Pickett explained to me that once I was adopted I'd share her last name. This troubled me because I identified with "Fisher," and without it I couldn't be Fish anymore. So I asked if it would be possible to keep my last name.

That was the last discussion ever to take place as it related to my adoption. Jill Edwards observed:

```
Although Mrs. Pickett has long in-
sisted that Antwone is no problem,
she is currently complaining that he
is "talking back to her" and is talk-
ing about having him removed. In
April, Mrs. Pickett stated that she
wanted to adopt Antwone. CW submit-
ted permanent custody summary to
Legal Department. Before a hearing
date was set, Mrs. Pickett had com-
pletely changed her mind. This was
partly due to the fact Antwone said
```

he wanted to keep his name, and
partly because she is stating she
cannot handle him at this point.

The "talking back" comment by Mizz Pickett referred
not to my general demeanor but to one comment on one
occasion that may have arisen from my frustration and
fear caused by the added pressure of living with mental
patients she brought home from the state hospital.

Hoping to fill every last bed in the house, Mizz Pickett
put a boarder in the room with me. I'm trying to go to
sleep and this dude is snoring at top volume. Mizz Pickett has a loud snore, too, and so does Mr. Pickett. But this
man is snorting and whining in his sleep two feet from
me. Plus, he smells like saliva. Just my luck, turns out his
big phobia is that he's afraid of water and refuses to
bathe.

The next boarder, Howard, is worse. Made of solid
muscle, Howard makes weird throaty sounds when he
speaks, his beard grows irregularly in clumps, and, being
used to the nuthouse (Mizz Pickett's term), sometimes
he'll lay spread-eagle in the front yard for hours. Mona
passes by one day and sees him, asking me later, "Who is
that dude in the grass?"

"I don't *even* want to talk about it," I say, and she
doesn't force the issue.

On Howard's first night, Mizz Pickett has him sleeping
in my room with me, and when I start for my bed at bedtime, I look in and see his ass is buck naked. Disgusted
and scared, I imagine the morning's headlines: "De-
ranged Nude Murderer Hacks Elderly Religious Couple
and Teen Ward to Death in Their Sleep." Here we have a
confirmed nut who can kill us all without warning and the
worst that's going to happen to him if he does is they'll

send him back to the hospital. Doing one fast U-turn, I veer right to Mizz Pickett's bedroom and tell her I don't want to sleep in there with Howard.

Her remedy is to give me a half a Valium and have me sleep on a cot in her bedroom, next to her snoring ass.

When he's not lying on the grass, Howard's other thing is to get up early in the morning, take a loaf of bread from the kitchen, and split. After several days of this, Mizz Pickett instructs me to get on my bike and follow Howard to see where he's going and what he's doing with the bread. Following her instructions to the letter, I trail him all the way downtown, watching him eat the bread, piece by piece, and dropping crumbs along the way—to find his way back to the Pickett house, I suppose. Of course, I resent having to waste my day doing this, but at least being on my bike is better than having to follow him on foot.

It's hard, though, to feel grateful for the bike, because I can't forget the resentment in Mizz Pickett's voice the time she made me carry the box with the unassembled bicycle in it up to the front porch. She gestured at it like an afterthought and said, "I didn't get-cha dis cuz I love ya; I got-choo dis cuz I thought-choo ought to have it."

Her sarcastic tone made me suspect that someone at social services thought I ought to have it. Though this was never proven, it left me with the feeling that the bike wasn't really mine, just on loan.

I stand by the bicycle, off to the side, watching Howard seated on a bench in the middle of a downtown Cleveland city square. He sits there, finishing his last crumbs of bread, and staring blankly off at the jealous pigeons. Having done my duty, I quietly slip away and pedal all the way home, where Mizz Pickett is mad because I've been gone so long. "Where you been?" she demands to know.

"Downtown . . ." I start to say.

"Downtown? Nigga, you went downtown?"

"Yes, ma'am, you told me to follow him. . . ."

She looks at me like I'm the crazy person, tossing me off with a "Boy, what's ill you?"

It was bad enough to know that no matter what I did to appease her, it would always fail. But what I found unbearable was when Mizz Pickett invented the ultimate taunt by calling me Howard all the time. And this is what causes me to make the one remark that might be seen as talking back. It comes at the height of my ninth grade academic successes, when our English teacher has been teaching us new vocabulary words—words like *loquacious,* such that we spend the rest of the week out of class calling each other loquacious and knowing what that means.

So one evening after an hour's derisive tongue-lashing, I look at Mizz Pickett and, without thinking, finally ask her, "Why do you have to make things so difficult on me?"

Stunned, she practically spits, "Difficult . . . difficult! Nigga, where're you learning them words like dat at?"

"School," I say, stating the obvious. Needless to say, Mizz Pickett won't take that as a defense for my heinous crime—not only of talking back, but worse, the sin I have committed of starting to grow up.

This was her perfect fuel for a frenzy of phone calls, each one beginning with, "Guess what that nigga did today?" It was also the basis for her complaint to social services that I had become unmanageable. Apparently, however, Jill Edwards didn't take her complaints at face value:

```
It seems that while another ward,
Dwight Perry, was in the home, Mrs.
```

```
Pickett couldn't say enough about
Antwone. She was constantly compar-
ing them and Antwone was always a
"good boy." Since Dwight's removal,
Mrs. Pickett claims she is having
the same problem with Antwone. She
states that he is lying and staying
out late and sassing her back. CW has
talked with Mrs. Pickett alone,
Antwone alone, and both together—
with no success in establishing bet-
ter communication. CW spoke with
Mrs. Pickett on the phone and asked
her to come in to the office for an
interview. Mrs. Pickett has been
most uncooperative and to date has
not made an appearance.
```

For the first time in my life, it was at this point, in December 1974, when I was fifteen, that I was put on record as to my feelings and observations:

```
Antwone complains that Mrs. Pickett
is inconsistent and unreasonable and
CW tends to agree with him. Although
Mrs. Pickett complains about Antwone
lying to her, it is apparent that
Mrs. Pickett does plenty of lying to
CW, of course, making it very diffi-
cult to work with her.
```

Though other caseworkers were able to pierce Mizz Pickett's facade, Jill Edwards seemed to have been getting the closest to the truth when she asserted:

Antwone is rapidly becoming a young
man and is beginning to feel the need
for some independence and decision
making. As with Dwight, this presents
a problem for Mrs. Pickett who expects
to be in complete control. Antwone's
continued placement in the Pickett
home is in a precarious position.

At the same time, my social worker also noted that
there were no known relatives of mine to take me in and
that other foster homes for teenage boys were unavail-
able. Therefore, she saw no alternative for me but to remain with the Picketts until I was eighteen. This may
have been different if she had been aware that my position was even more precarious than she thought.

She didn't know, for example, that there had been a
boarder living at the Picketts for whom I was made to be
a caretaker and orderly—which, of course, included
bathing the grown man, a duty I hated and performed
with utter resentment.

Not everything that went on with the boarders made
me miserable. Some of them were interesting and smart,
in spite of their peculiarities. Sometimes, there were two
or three boarders staying in the house at one time, lend-
ing to the everyday gloom a festive, circuslike atmosphere. There was Little Robert, a classical music lover, a
short man, not to be confused with Big Robert, a comical
guy the size and build of a professional heavyweight
boxer. He worked as a janitor at the airport and on his first
day at work, Mizz Pickett made a dash for the room to inspect his things—only to discover that Big Robert had a
closetful of ladies' dresses and shoes and women's hy-
giene products.

So at breakfast the next morning Mizz Pickett goes all whoop, whoop, whoop, getting in Big Robert's face, asking why he has all that shit in his closet.

"Because," says he, "I'm getting married."

"Oh?" says she, relieved to hear it. "That's niiice. I didn't know you had a girlfriend."

He answers, "I don't. But when I find one, I'll be ready."

Another day, we're in the hall upstairs—me, Big Robert, Mizz Pickett, and Lizzie (the evil Pickett twin, now grown and married with three kids)—when something causes Mizz Pickett to start with the whoop-whoop again on Big Robert.

"Nigga," she calls him, giving him the old hand-on-hip lean and wagging finger.

"Nigga?" says Big Robert, and he hurls it back to her, saying, "You a nigga."

Everyone freezes. Never since the day I was brought to the Picketts have I ever heard anybody call her that. Mizz Pickett is thunderstruck, the very breath taken out of her, hearing herself called a name that probably no one has called her since her childhood.

Trying to hide the smile of glory breaking out on my face, I start to back into my room as Lizzie shouts, "Don't be callin' my momma no nigga!"

"Well, she called me one," he responds indignantly.

In the privacy of my bedroom, I trip in disbelief. He called her a nigga, I keep repeating to myself with amazement.

Big Robert may have been a little touched in the head, but what he had just done made me think him a philosopher king. And he became, for that moment, my hero and my redeemer, giving voice to a concept that had been building up inside me until now—that anyone who demeans another person in word or deed demeans himself or herself.

Unfortunately, Big Robert's great lesson was lost on Mizz Pickett, and she got rid of his ass within the hour. I was sorry to see him go, nutty as he was. The boarder I liked best was Benny. An attractive man, he had problems, possibly stemming from having been in Vietnam, but he was cool. Just this dude trying to keep the world from tearing him down. So he drank, not at the Picketts, but over at some bar on 105.

Mizz Pickett had been making noises about Benny stealing from her, but had never been able to finger him. Well, it's a sunny Saturday morning in June, the temperature rising too fast for this early in the day, and I'm standing on the street corner with my friend Michael Shields, minding our own affairs, when Mizz Pickett pulls up in her green Catalina.

My first thought is, What am I in trouble for now? My next thought is how I'm going to keep Michael from seeing what a kook Mizz Pickett is. The only person who knows I'm a foster kid is Fat Kenny, but that's all he knows. Even though I've never referred to her as my mother, I never refer to her at all. But lately, I've been getting questions from the others about my circumstances, which up till now I've done a masterful job at hiding. Michael, picking up on the fact that Mr. Pickett is so old, has recently asked, "Is he your father or your grandfather?"—a question I dodged by changing the subject. And none of them, except Jessie, who doesn't give a damn, has ever been inside the Picketts' house.

Before she can get out of the car, I run to the open window on the passenger side and lean in.

Steaming mad, Mizz Pickett says, "Benny went in my purse and got my money. You know where he at?"

"No, ma'am."

"C'mon, get in this car, we gone go find him."

Since Mizz Pickett has no idea which bar Benny frequents, her plan is to go and check every place on 105 until she apprehends him. It turns out, she has no evidence against him except for the fact that since she hasn't given him his weekly check (she keeps the boarders' allowance checks just like mine), he wouldn't have left the house if he hadn't stolen money.

In the passenger seat, I sit and listen to her fussing about Benny, keeping one eye on her and the other on the road as we careen through traffic. Mizz Pickett still juts her head over the steering wheel and, in all this time, her glasses have still never been able to stay put. Now that I know how to drive, I'm aware of how dangerous her driving really is sometimes.

My learning to drive was not the result of any generous schooling on the part of the Picketts. It had happened by accident when I was as young as twelve and found that washing the car for Mizz Pickett without being asked would usually win me one of her bones of approval. In the cleaning process, I learned to wash one side of the car, then drive it down to the street, turn it around and pull back into the driveway so I could clean the other side. Before long, Mizz Pickett began sending me out on winter mornings to warm up the car and scrape the ice off the windows. Pretty soon after that, she was asking me to chauffeur her places when she wanted the car returned home for someone else.

On one such occasion, the temptation and the opportunity to share my driving talents with my friends were both far too great to turn down. Cruising down Pasadena behind the wheel of that car with three of my friends in tow was so thrilling, I had to do it one better and take us to the big time, 105. Waving to other friends from school, I was having the joyride of my life until it hit me like a knockout punch that we were going to pass right by Lizzie's husband's gas station. At the last second, I turned right around and headed back to the Picketts' house.

Mizz Pickett eventually found out and promised that I would never get behind the wheel of her car again. It was all right to give me adult duties, like driving her, but not adult desires, like driving myself and my friends.

"Stay here," she tells me as she pulls into a parking space in front of the first bar we're checking for Benny. With that self-congratulating look of outrage, Mizz Pickett marches into the bar, disappearing into its smoky darkness for more than ten minutes. When she emerges, there is no Benny being pulled by the ear or led to her car in handcuffs. But Mizz Pickett is grinning slightly, for reasons she doesn't decide to share.

The same thing happens at our second stop, only this time she's grinning even more. And again at the third bar. By the time Mizz Pickett comes out of the last bar, still with no sign of Benny—he is long forgotten by now—she is giddy as a schoolgirl.

Mizz Pickett slides onto the seat and turns to me with a sigh. She removes her glasses and folds them into her lap, and the soft look on her face tells me I'm about to hear a confession. "I know there better be a God," she begins wistfully, "much as I been sacrificin'. Bein' good, prayin', waitin' on the Lawd and what-not . . . I could be out here havin' fun. . . . You hear me . . . fun!"

In two blinks of the eye, her life flashes before me as it could have been. For this instant of revelation, I understand her. I understand that for the first time in a long time, she may have known the taste of wine, hummed and swayed openly to the down-home blues in that string of dim, musty bars, as a search for Benny reacquainted her with a self she had long ago set aside. Probably she went in there to one of those bars and some drunk-ass dude may have said something like, "Hey, momma, what's poppin'?" and it transformed the mundane black-and-white docudrama of her mean old life into a living color romance. Fun. If it wasn't for her religion, she would choose to go out right now in a blaze of lust-filled, Devilish glory.

The moment of understanding was only a passing one. Mizz Pickett quickly went back to her life of sacrifice and waiting on her mansion in the heavenly skies. She was pretty much as mean as she had been before, though maybe not as relentless, simply because she was getting old and tired.

This fact, unknown to me, was to figure prominently in my murky future. Without telling me anything, the Picketts were looking for ways to close down shop and pull up stakes. Mizz Pickett began to disappear for months at a time, for what I later understood to be exploratory trips down south—where, amid relatives still in the area, she must have been contemplating retirement.

Mr. Pickett was also away much of the time, or locked away in his room. I was left in the care of Lizzie. Motherhood and life hadn't softened her in the least. Moving into the house with her three kids, she took over Mizz Pickett's role as head warden like she'd been in training for it her whole life. She must have stored up all her resentment against me for invading her house when she was a teenager and had returned to seek vengeance.

One of Lizzie's more reprehensible acts was when she asked to borrow my record collection for a party she was hosting. When it came time to return my prized possessions to me, she decided, in typical Pickett fashion of justice, that she flat out wasn't giving them back. That killed me.

It's the dead of winter in early 1975, the middle of the ninth grade. At fifteen, I'm tall, busting out of everything I own, and there's nothing that can be done to repair my worn-out canvas sneakers. Walking to school in the snow with Sonya one morning, she says, "Fish, when are you going to get some snow boots?" She meant the black rubber boots that went over our shoes, the kind that clipped up tight to your shins to keep out the wet.

Sonya and Fat Kenny live with their single working mother, a woman devoted to their well-being in every respect. I wonder if they even know how fortunate they are. But I also know they have no point of reference to understand my situation, so I shrug off her question—just like I'm trying to shrug off the icy wetness seeping through my shoes' fabric and old rubber soles. When we arrive at FDR, I hurry off to find the nearest radiator, hoping to dry my shoes as much as possible before my first class.

By spring, the bottoms of the shoes have permanently hardened into slippery plastic, such that when I run through the halls trying to get to class, my head turned to another student as I crack jokes, I don't see the wall I'm about to hit and when I do, my shoes have no grip. Trying to cushion my collision with the wall, I extend my arms in front of me, bending my left elbow backward to the point of breaking. After a large group of students help me to the school nurse, during which I start to go into shock, I am sent to nearby Forest City Hospital, where I was taken the year before when I broke the wrist of the same arm at school. There it is explained that I have a separated funny bone and a fracture of my elbow and that I should be taken to Rainbows, Babies, and Children at University Hospital for immediate care.

Lizzie didn't take me that day, nor the next. On the third day, in the evening, she did not take me herself but asked her father, Mr. Pickett, if he would take me. In these three days, I had been sent to school, enduring fever, throbbing pain, and continued swelling. When Mr. Pickett got me to the hospital that night, he was informed that I would need intricate surgery involving two pins to reattach my funny bone. But first, because of the swelling, I had to remain in the hospital for three days with my arm elevated.

There I am, daydreaming, lying in my hospital bed, when I glance up through the glass to the nurses' station. I see Mizz Pickett out there, in that same old tired-ass coat with that tired-ass fox on the collar with the same tired-ass beady eyes, still staring at me after all these damn years, like a member of their family. Later, I'll wonder if he had actually been a meal at one of their finer occasions.

As if in slow motion, Mizz Pickett seems to sense my eyes upon her, and she turns to me like Bela Lugosi, casting her burning eyes back at me. Long ago I gave up the notion that she was an alien monster. I take that shit back.

There are six beds in the room and as she comes toward me to sit at my bedside, the other children grow quiet, as any child would when a vampire enters a room.

"Well. I see you broke your arm again," she says, her voice shaking to hold back her anger, I assume, for having to rush back to Cleveland on my account.

"Yes, ma'am." Without warning, I begin to cry.

"What's the matter?" she asks.

To my surprise, I tell her the truth: "I'm lonely." It's not because I think she'll understand, but more to save myself from drowning in swallowed hurt, as if the articulation of my loneliness has slipped out of me of its own accord like a fugitive giving up his hiding place. Still crying, I continue, "Keith is gone, Dwight is gone, and Flo is gone, and I don't have nobody. I don't have nothing."

"You got Lizzie."

I think, After twelve and a half years, she doesn't know her daughter hates my guts? Or maybe she's pretending not to know.

In any event, Mizz Pickett chooses this moment to make her exit, standing first and saying, "Well. Since you cryin', I'm gone go." With that, she turns away and, through the blur of my tears, I watch her and her fox leave the room.

Over the next six months, Mizz Pickett continued to go out of town periodically, with Lizzie continuing to stay in the house when she was gone. During this time, I was given physical therapy and regular follow-up medical attention. According to my file, Jill Edwards never knew about the Picketts' initial delay in getting me to the hospital, nor was she fully aware of Mizz Pickett's frequent absences. The episode, however, was enough of a red flag that my social worker again insisted on a one-on-one interview with Mizz Pickett, followed by one with me, after which she wrote:

Foster mother is often difficult to deal with. She is likeable, but very manipulative and inconsistent. She will praise Antwone during one contact and ask for his removal at the next. Although Antwone is very used to her, he has become less willing to

put up with her and is beginning to
stand up for his rights and opinions.
This is very irritating to Mrs. Pickett . . . because she always wants
her children to be grateful.

Ms. Edwards described how Mizz Pickett had previ-
ously announced she was moving to the South (something I never knew) and that she would be happy to take
me, but "Antwone refuses to go." Then, my social worker
reported:

In the next few months, Mrs. Pickett
changed her mind several times. She
finally stated that Antwone was part
of the family and will always have
his place with them.

Therefore, she wrote, in conclusion:

The casework plan is for Antwone to
remain in the Pickett foster home
even though it is not an ideal situation. Removing Antwone at this
time would be very detrimental as
his identity is with this family.

While Ms. Edwards had been misled by Mizz Pickett
on several points, she was probably right on this last note.
By this time, my coping skills were so attuned to survival
in the Pickett penitentiary that removing me to any other
environment would be akin to taking an animal out of
captivity and releasing him into the wild. But that, in fact,
was exactly what was about to happen.

The beginning of the end came in late 1975, around the time that Jill Edwards transferred my case to a new social worker, Patricia Nees. It was some months before I spoke with her, but when I did, Ms. Nees—another warm, sincere young woman—relayed Ms. Edwards's personal good-bye to me, explaining that Jill had left the agency because she was getting married. This made me feel that Jill Edwards was personally concerned about me and had alerted Patricia Nees to some of the problems I was facing.

A reference to the reason it took so long to meet Ms. Nees was made in her first entry:

```
From the time caseworker was as-
signed this case . . . for a period
of three months, caseworker was un-
able to reach the foster mother,
Mrs. Ulysses Pickett, or our ward
Antwone Fisher.
```

That period in question was marked not only by one of Mizz Pickett's longest absences, but by the presence of this group of marauding invaders who appeared to be a family that just moved in one day with their furniture to the Picketts' house. Mr. Pickett was possibly buried in his room; or maybe out of town. Whatever the case, nobody offered any explanation when I arrived home from school and saw that, with the exception of Mizz Pickett's padlocked room, all of the bedrooms upstairs, including mine, had been taken over. There was a lady and several teenage kids, none of whom seemed to notice or care that I was living there. The only kid in the house whose name I even knew was a lanky teenager named James Tally. Knowing of his reputation as a troublemaker, I had my doubts about all of them; then again, I assumed that Mizz Pickett and the lady were friendly.

Over the past few years, my enjoyment of school and my group of friends had helped me keep my wits about me when it came time to don my battle gear and go home. But I was at a different school now, and things weren't the same. Most of my close friends—Freda, Mona, Jessie—were going to Glenville High; where I lived was the dividing line that meant I went to John Hay High School. Plus, the uncertainty of my day-to-day existence made me wonder why I should bother working hard in my classes. Some days I went to school and cut class; other days I didn't go at all.

In the meantime, the walls of illusion separating my two worlds were crumbling. I must have told someone about James Tally at my house, maybe Fat Kenny, because Sonya came to me one day wanting to know, "Hey, Fish, why is James Tally livin' at your house?" This wasn't like the shoes, when I could change the subject. This was crazy shit.

Before, my friends looked at me, and I seemed relatively normal. Not anymore. Everything was getting raggedy—first with the crazy boarders and then no winter clothes and now an unsavory family that came and went at all hours and ate at all different times of the day, never together.

Since no one included me in their meals, nor appeared to be looking after me, I was left to fend for myself. Occasionally, the lady gave me some change, and I would go to the store for chips and a soda for myself. Once, after two days without eating, I scrounged around in the pantry and found an old potato.

Putting it in the oven to bake, I was so famished, I couldn't wait till it was cooked and ended up eating the potato half raw. In desperation, I called Mercy, who seemed sympathetic but still leery about doing anything. It was then and only then that it dawned on me what Dwight had understood years before. I wasn't a Pickett and would never be. If I had been a part of the family, it would have been arranged for me to stay with any one of them, not left with strangers without an iota of an explanation in the middle of winter.

As inexplicably as they had come, suddenly the James Tally gang prepared to leave, with a flurry of cleaning of scorched pans and repairs of broken household items amid murmurs that Mizz Pickett had called from down south and was getting on the highway that day. They left only a few days after Mizz Pickett returned.

The next time I saw James Tally was at a friend's house. As I was there to buy an old stereo receiver, I had come with all the money I'd earned after a recent snowstorm gave me an opportunity to hire myself out shoveling snow for neighbors. When I saw James, I greeted him courteously. He, on the other hand, approached me calmly, punched me in the stomach, grabbed my collar, and then reached into my pocket and took the entire contents—all thirty dollars.

"This is for the trick y'all played on my momma," James Tally proclaimed, waving my money in the air, before running out of the house. Afterward, I wondered if he knew ahead of time that I was going to be there; if he did, it meant that my friend set me up.

Whether that was so, I never knew. And I never knew what it was that Mizz Pickett had done to James Tally's mother; nor would I dare to ask Mu-deah. With her return, the house was almost back to the way it had been before. She was there, running the show like the ringmaster she was, and Mr. Pickett was there, too, with his mug of Kool-Aid at dinner. The clutter and the chaos of the past many months seemed swept up for the most part as an uneasy calm settled over the household.

Beneath that calm, I felt numb. Whatever crumbs of hope I'd been scattering through the snows of my life, believing that someday I might be led back to my real home, were now lost completely. And then the rains came, bringing with them a premature spring, turning everything to cold, gray mush as I fell further and further into the hopelessness of my reality.

The end comes the morning after an incident having to do with Jessie. This is a morning unlike any other I have known at the Picketts' house. In that way that I've come to determine what kind of day it's going to be by the smell in the air, I'm also attuned to the morning sounds. On any other day, when I'm up early to iron my clothes before school, the house has a medley of snores. Mr. Pickett has his snore. Mizz Pickett has a snore. Laddie has a snore. The house itself has a snore.

But this morning, there are no snores. No sounds. Just quiet. And something brand-new—no smell at all. Even the rain that's sure to fall soon comes without an odor. Outside the house, small splashes of light from streetlamps peek through the mist, like they're trying to cheer me or distract me from thinking about what I know is a terrible storm on the way.

Tired of having to stay indoors, or just looking for something to do, I planned the previous evening to go to the movies with Jessie. We hadn't been hanging out much, partly because he was going to Glenville and I was at John Hay, and partly because I was keeping to myself more often than not. So when he asked me to go to the

Hippodrome, the movie theater on 105, and said he'd pay since I didn't have the money, I was glad to take him up on the offer. At no time did I anticipate any problems.

That is, until late that afternoon when Jessie decided to come get me by sauntering up to Mizz Pickett's sacred front door and ringing the bell, knowing good and well that his ass wasn't supposed to come to the front door. Nobody was supposed to come to the front door. Like everybody else, Jessie knew that he was supposed to come to the side door. Or, better yet, he could have waited for me on the sidewalk or at the movie theater and I could have left unnoticed through the side door.

Instead, I had to wait in my room upstairs, pacing the floor, just imagining Mizz Pickett opening the door to Jessie. I visualized him standing there, hands in his coat pockets, with that permanent scowl on his face, undaunted by her sneer and unimpressed by her full-figured aging form in her everyday housedress, underneath which her stockings were rolled up to the knee, as she folded her arms underneath her huge bosoms. She never liked Jessie, a brash kid who didn't give a damn about rules, hers or anyone else's, who stopped by asking for me whenever he pleased, and whom she probably had seen high up on 105 or dancing on the fire escape at Parkwood Elementary. She deemed him to be "that hoodlum nigga Jessie, yo' friend," which was how she referred to him.

After I heard the front door slam, certain by its sound that she had sent him away, her voice sliced through the air, "Nigga, get down here."

Since I was the only one left, the Last of the Mohicans, I knew she was talking to me. Girding my loins, setting my jaw, I went downstairs, resolved to be stoic and accommodating, or whatever I had to do to get through whatever she was going to do. Later, much later, I would conclude that nothing I could have done would have placated her. Indeed, I came to believe that she had been preparing for this, almost as an opportunity—that the next time I got into trouble with her she was going to unburden herself of me and she had a script ready. But not knowing that, as I went downstairs, I was fairly confident with my usual arsenal of "I'm sorry," and "Yes, ma'am," and "I wasn't thinking."

She started in immediately with, "That was yo' friend, that hoodlum nigga Jessie; and I sent him on!"

"Yes, ma'am."

"Don't you sass me, ya hear?" Mizz Pickett pulled out her well-worn lecture about the evils of the movies, reminding me, "That the Devil's carryin' on! And whatchoo know about that movie house up there? Hangin' out with that hoodlum Jessie's gonna get-choo in a world of trouble. Mark my word, you rotten hardhead nigga. I told you once, I told you time and time again, stay away from that nigga you like so much!"

"Yes, ma'am," I repeated, staying cool, trying to sound my most sincere. "I wasn't thinking."

Not hearing anything, she raised her rant, "You jus' won't mind—"

"I'm sorry," I ventured.

"Don't talk back to me!" she shouted, and went on, and on, until I tuned out what she was saying and said nothing, standing there impassively until she spent her wind and her words. Then, I silently turned, head lowered, naturally, went back to my room, closed the door behind me, and sat down on the edge of my bed. Resting my eyes for a second, I felt that all things considered, I had survived.

At that second, the door flew open and she stormed in. "Don't you slam this door!" she began.

I stood up from the bed and protested, "I didn't slam the door, I was just closing it—"

That did it. Before I could finish, volts of electricity appeared to ignite her rampage and she bent down, whipped off her shoe, a leather slip-on with a small, square heel, lurching at me with it and hitting me—*wham!*—on my neck. Reacting automatically, I snatched it from her. Mizz Pickett stood back in spiteful, speechless amazement.

She looked at me for a handful of long, very quiet moments, followed by a firm pronouncement and a pointing of her index finger: "I wont-choo outa my house."

The finger began to swing every which way, as she repeated, louder and angrier, "I wont-choo outa my house," over and over, in the middle of which I did say, "I want to go!" But she kept up her chant until it seemed the pendulum of her finger had pointed to every subatomic particle of me.

When she finished her tirade and left the room, I went to the door and closed it, figuring the worst was over. She'd threatened me a million times with promises of, "I'm gone send you back there and let the white folks deal wid'-choo," usually of late as she compared me to one of the crazy boarders she kept and called me one of their names, usually Howard.

Whether or not I had cause to be worried, I couldn't decide by my own feelings. They had become desensitized, and I was too beat down to separate juice from pulp anymore, so that I was the same mush that I saw around me. By now, I didn't know how to feel, maybe because when you step on mush, it's too mushy to feel it's been stepped on.

Dinnertime came and went and I stayed in my room. It was my habit to spend hours in here anyway, listening to my cassette player playing songs I'd taped off the radio, drawing, being still, solving stuff in my mind, and riding away on my magic carpet of thought. My plan now was to avoid her for the rest of the night and not be around her in the morning, by slipping out before she was up. If I went to school and made it back in the evening before my usual time, I could maybe surprise her by straightening up or cleaning.

As I strategized, I could hear Mizz Pickett downstairs on the telephone, making phone call after phone call, announcing to each person, "I'm sending that nigga back." At first, it was all too reminiscent of a hundred other times when she threatened to send me back. But as the night wore on, I did begin to think she was saying it too emphatically and telling too many people, expressing it not as a threat but as an action she was about to take.

Later on, I realized she was probably trying to brace everybody for the news. True, nobody who inhabited the Pickett universe was really invested in me. But for her to announce without prologue that, after so many years, Antwone was gone might cause undue scrutiny—or so she may have guessed. For her it was much easier to make it be about something the nigga had done. And this, later on, I would find the most reprehensible of all: that at the end she chose to do what she did, the way she did it, to save face. Such a foolish proud woman, she probably felt her good name would be ruined if she sent me back without a cause, without a big stink. But in the course of human decency, it wouldn't have cost anything to explain to family and friends, to social services, and to me: We're getting older, and we were already old when we took Antwone, and we had him all these years, and he's a good kid, but unfortunately we've got to get out of the foster care business and go back to our small town in the rural

South, where we're from; it wouldn't be right to take Antwone out of high school and away from his friends.

This would be my feeling only in the time to come. It's true that the Picketts had been talking about the joyful day of going home for as long as I could remember. In fact, when I was seven Mizz Pickett had taken me and Dwight there for a brief visit. Unknown to me, she had been told by Child Welfare that removing us to any other state even temporarily was forbidden. At sixteen I was still not sophisticated enough to grapple with the legalities. I didn't understand that if they were going to live in another state, and I was a ward of the state of Ohio, I couldn't go unless they adopted me. Not that I wanted to go, because Cleveland, Ohio, as far as I was concerned, was my only real home.

So it didn't occur to me that there were issues much larger than her being unhappy about Jessie and me going to the movies. Therefore, in the thick of it, even as I heard her on the telephone that night much later than usual, I still held to the notion that she would cool down by morning. As in the past.

But the past is just that, and this morning, this present, has no smell and no sound. It means she's up, not cooking, not waging any other war on dirt or mess. She's down there, lying in wait. Now last night's phone conversations echo in my ears as I begin to feel like maybe she ain't messing around this time. So instead of my plan to slip out, I decide to be slow about leaving.

Waiting and listening, I allow breakfast time to pass. After I dress for school and start down the steps, that mush inside me begins to churn with a mixture of battling emotions. Buried down deep is my desire to leave, from which a cockeyed hope springs that she isn't kidding. The rest of me pulls in the opposite direction; my fears tell me to cling to the shipwreck island because it's all I know, because I don't want to hope that it can be better somewhere else—like Keith and Dwight did—only to encounter new enemies in new lands.

With all that bubbling inside, I enter the kitchen and see Mizz Pickett seated at the table, just waiting, and I say, "Good morning," as is the required greeting of the early day.

"Is your junk packed?" she asks, voice steady and to the point, pinging me with the realization that she is serious.

"Yes," I lie.

"Bring it down, cuz you goin' back this morning." She mutters something about one of her nephews being on his way to drive me.

Going back? I turn the phrase over in my head, looking for clues to a question I've never answered: Where is "back"?

Upstairs I go as a new feeling sparks inside the mush. Anger. It's a feeling that's been there a long while, sleeping uneasily. Suddenly, I feel it wake up, like an oven that's been lit.

Following shortly after me, Mizz Pickett appears at the door holding two folded A&P brown paper grocery bags she has ready. She tosses them into the room, as if she's skimming stones on the water, telling me, "Put ya mess in these. I don't wont-choo takin' none of mi' niiice suitcases."

If this is it, I think, I better decide what I'm going to take with me. But as I look around the room, in the closet and in the drawers, I don't see shit that I want. Into one of the bags goes a pair of pants and some underwear. That's it. For the rest, I don't want any remembrances. Even my prized possession, the cassette tape player, will remain behind.

(Even later, when Mizz Pickett will deliver all of my

things to social services, I refuse to take them. Of course, she will deliver everything except the tape player.)

Though it's already begun to pour outside, Mizz Pickett apparently changes her mind about having her nephew drive me, indicating as much when she comes to the door and tosses bus-fare money at me, with that same skimming-stones-over-water throw. "Catch the bus," are the last words she speaks before going downstairs.

In my coat, with bus fare in my pocket, a rolled up A&P grocery bag under one arm, I pass by her at the stove where she is cooking breakfast. She says nothing. I say nothing. Then I open the kitchen door, step outside into the rain that falls steadily on and around me, close the door behind me, and hurry away to catch the same bus that normally takes me up 105 to John Hay.

That churning mush inside me has changed into a turbulent volcano. There is sadness, almost despair in knowing Mizz Pickett isn't going to come running out after me in this rain to retract her decision. There is that heat of anger rising, like a gush of lava that's been waiting for centuries to burst forth, destroying everything in its path. And to my surprise there is a new feeling that I encounter with disbelief: happiness. I'm so happy, a part of me is leaping like a damn deer in the fucking woods.

By the time I board the bus, the rain is beating down in huge drops that soon turn to sheets of water that make it hard to see the sights outside the windows, some of which I will be seeing for the last time. When the bus nears Mt. Sinai Hospital, I pay close attention, taking in every detail of a place that has held such fascination for me all these years for reasons I've never known. Over the past several months, I've looked at it this way every morning and every afternoon, never once failing to turn my head when we passed by.

The seats are all taken, so I stand in the aisle, too nervous to sit anyway, letting the sounds of the bus and the thumping rhythm of the rain match what I imagine is the machine of my mind, with its complex of pistons pumping every which way. I'm thinking about my friends a lot, and about myself even more.

With all credit to Mizz Pickett, it takes this one bus ride to empty me of all the self-confidence instilled in me by Mrs. Profit, which had flourished on into junior high. I can practically feel it seeping from me. Outwardly, of course, I'm no different than any of the other kids on their way to school, who are mixed in with others this morning on their way to work. The only difference is that at Euclid when the John Hay students get out to transfer to the bus across the street, I get out and stay on the same side and use my transfer for the bus that goes downtown.

I disembark near the Child Welfare social services offices, across the street from the TV station, Channel 5. In the rain, the brown brick state buildings seem unaffected, sturdy in their bland institutional way. This is "back," I think, as I push through the familiar lobby door.

The door having just been unlocked, the lobby is empty when I enter, except for the receptionist. Noting my distress, she asks, "How can I help you?" which I read as "What are you doing here?"

"She sent me back," I say.

"Who?"

"Mizz Pickett, she sent me back."

The receptionist tells me to have a seat as she goes to call my caseworker. Moments later, she reappears with Patricia Nees, who introduces herself. This is our first meeting.

Though young, Ms. Nees has a maturity and seriousness I detect right away as she sits down next to me in the lobby and asks what happened. She also has a softness that makes me feel, even though I don't know her, that she's on my side.

As I begin to talk, people accumulate in the lobby while the business day gets under way. In the middle of my explanation of the events that have led me here, Mizz Pickett makes her dramatic entrance. Now all the dots are starting to connect. Now I begin to see that she has had a plan all along but that the rain has foiled it. Clearly, she didn't want to drive me here because she didn't want to face a jury in person. But now she can't let them think she was so callous as to send me out in the rain. So here she is for a little damage control.

Wearing one of her finer winter raincoats and a clear plastic rain kerchief tied delicately under her chin—like her ass is Dorothy Dandridge—she is ready for her show, right here in the lobby of social services, her favorite place to cut up.

But before she can launch into her speech, Ms. Nees stands and asks why she's been unable to make contact for three months. Thrown off track for a moment, Mizz Pickett tells her tale of living in the South during that time while her responsible daughter was living with me as my guardian. Skeptical, Ms. Nees tries to question her, but Mizz Pickett interrupts, leveling her pointed finger toward me, saying, "He took my money. I'm down here to get my money."

"Wait a minute, Mrs. Pickett," says Ms. Nees calmly. "What money?"

Mizz Pickett pats her beloved patent-leather purse that hangs so girlishly from her wrist, calling me a thieving, lying, scheming boy who can't be trusted. "I tole him he couldn't go to the movies, and he cussed me and swung his fist at me," she announces to the room. She recounts the tale that I missed hitting her but it scared her so much she was forced to have me removed and that when she told me to go, I robbed her.

The heat inside me flies so high it burns my ears and propels me from my seat and I start pulling and turning out every pocket on me, showing their abject emptiness. Everybody in the lobby—my social worker, the receptionist, all the strangers—everybody has focused in on our scene, and I can feel their eyes on me, but I don't care. I don't care because of the revelation that Mizz Pickett has no power down here, that she is so far out of her element she can't do anything to me. I kick off my shoes and am getting ready to pull off my socks when Ms. Nees says, "Antwone, you don't have to do this. Put your shoes back on, put your pockets back in."

Then I speak, shocking everyone, including me. "I don't have her money, and she knows I don't have her money." And I go on, real articulate, too, in a kind of out-of-body state as my thoughts become crystal clear, needing to be heard, such that my brain takes over with my mouth moving in accordance. "She's always accused me of stealing," I tell Ms. Nees. Then I turn to Mizz Pickett and accuse her: "And you are lying."

Never before have I said that curse word of all curse words to her, the devout churchgoer. To this, I comment out loud, "And you call yourself a Christian lady, but that doesn't mean anything because we both know you are lying."

Everyone in the waiting area seems to draw in a collective gasp. Ms. Nees is looking at me, the receptionist is looking at me, Mizz Pickett is looking at me—each with her own degree of shock. Then Mizz Pickett's face suddenly changes as the twisted pinch of her features relaxes and she drops the mask of self-righteous indignation she has prepared for this play, melting her expression to a softer mask I've never seen before. She doesn't say a word, but the look she wears now is intended to say much more. Part of the softness is contrived, like the witches in all those cartoons trying to look innocent when they smile and act surprised that anyone could accuse an old, helpless, humpback woman of eating children.

In Mizz Pickett's look, I also see her realization that the final act is over, and I can just hear her fussing at herself: *Well. You got in your car and you came down here and now he's embarrassed you. You shoulda left it well enough alone.* There is something else in her look that I can hear her saying to me: *Well. I'm going home to my warm house and you don't know where you're going, do ya? You made me look bad down here, but I got what I wanted, didn't I? I'm free of you.* All in all, it's a self-satisfied crazy look of a woman whose only real concern at this moment is how she's going to get the hell out of the lobby.

Throughout my speech and her look of response, I have stood in the same spot, my pockets still turned out and my shoes off. Ms. Nees now repeats quietly, "Antwone, put your shoes on and put your pockets back in. Please sit down." Louder and colder, she says to Mizz Pickett, "Since you two can't seem to resolve this, we could have him stay with your daughter, as you discussed with Jill Edwards."

What daughter? Realizing none of the others would offer, it must be Mercy. Suddenly I don't feel abandoned.

That is, until Mizz Pickett wrinkles up her nose with total disgust and says to Ms. Nees, "After what he done to me?"

The ghost of Dwight sits down next to me, puts his arm around my shoulder, and whispers, *Do you get it now?*

Now I get it. I never had a family. *Yeah,* says the ghost, as he vanishes.

Ms. Nees gets it, too, that whatever plan they had for me has just been withdrawn. "Very well," she says to Mizz Pickett, nodding to the door, "unless there's something else you want to say, there's no need for you to stay."

Without blinking, Mizz Pickett turns, gets on her broom, and flies away.

For her entry in my file, Patricia Nees described the events of the morning of March 12, 1976, very much as I remembered them. Of course, she may not have known how triumphant it felt to see Tha' Lady Isabella Pickett finally shut up—having been shut up. She may not have known the intensity of the despair that was welling up inside of me as she led me out of the lobby to her workstation, a small cubicle next to a window, through which I could see that the rain hadn't let up in the slightest.

But for all that she didn't know, Ms. Nees would prove over the next hours to know my needs better than I myself did. She began by taking out my file. Before opening it, she asked me questions about myself—what my interests and skills were, what I enjoyed doing, what I didn't like. At first, I responded in the manner that survival in the Pickett world had trained me to do—playing young, playing dumb:

```
When this caseworker asked him about
himself, Antwone said, "I am shy and
not good with words. I like Art and
watching basketball. I work hard if
```

```
I'm interested in my job. I don't go
to school and sometimes I am just
lazy."
```

That was the sort of self-deprecating shuffle that had become habit. But in the next breath, I spoke frankly in a way that was out of character:

```
I hate Mrs. Pickett and all that she
stands for because she won't let me
have a life outside of her. Anyhow,
she doesn't really care about me.
I'm sick of being told I'm no good.
```

In the course of talking to me that morning, Ms. Nees introduced me to her co-workers and her supervisor, an energetic good-natured man in his early forties. Like Patricia Nees, her supervisor and everyone treated me with sincere concern and, not only that, with interest in what I wanted or cared about. Later, after deciding to place me temporarily at Metzenbaum Children's Center, the orphanage where I had begun life, Ms. Nees's supervisor and the rest of the staff sent me a gift—a basket with an assortment of men's cologne in it. There was Brut, Canoe, English Leather, all the popular brands. The gift alone was wonderful, but the fact that they understood how important it was for me to be acknowledged as a young man, not a boy, was exhilarating.

After such a kind reception, when I was sitting beside Ms. Nees in her cubicle that first morning, with the sound of the rain behind me, she brought me back to reality, admitting, "Antwone, I don't know where we're going to place you."

My thought was, Damn, I'm gonna live right here in the office.

Ms. Nees brought up Boys Town as a possibility, later
noting my feelings in my file:

```
Antwone was vehemently opposed for
two reasons. 1. Glenville was the
neighborhood he had grown up in for
13¹⁻₂ years. He had poor memories of
this time and had not yet dealt with
his separation and rejection. 2. His
foster brother, Dwight, was placed
at Boys Town and came back with neg-
ative stories of the boys. . . .
Antwone did not like their "gangster
like attitudes."
```

Although my feelings existed, Ms. Nees's report went
on to say that I agreed the following week to go for a pre-
placement interview, at which time I made my position
clear to the program director.

```
The director allowed Antwone to
choose, feeling that there was no
sense placing him there if he was so
strongly opposed. The director and
CW agreed we did not want to make a
"runner" out of him if we could help
it.
```

What was most memorable about the interview at Boys
Town was what happened afterward, when I was on my
way to the parking lot with my social worker and I looked
up to see Mizz Pickett standing with folded arms, watching me. She was near her car—why she was there, I
didn't know. She gave me the same look of "Well." It was

that villainous face she melted into in the lobby of the social services offices—soft yet sinister, and now tinged with satisfaction that my ending up at Boys Town proved true every pronouncement she'd ever made about my worthlessness.

And that was the last time I ever saw her. The sight of her tore into me, confirming every instinct inside to get away as far as I could from her, away from the neighborhood, away from Glenville, away from Cleveland.

Observing me then, Ms. Nees must have understood why earlier I refused to have any of the clothes that Mizz Pickett had delivered to social services. So Patricia Nees pulled a few strings on my behalf and procured a hundred dollars from social services with which to take me downtown to go clothes shopping at Higbee's. It was a spectacular afternoon, not only because this was the same department store where years before I had accompanied Keith when his mother took us shopping to buy him clothes, but also because I had never in my life had so many new things at once. In my past, I had been given a new pair of pants and a new shirt at a time, but this was five pairs of pants, five T-shirts, underwear and socks for days. And there was more—I got to pick everything out myself. Ms. Nees's comments were all compliments: "That looks nice on you" and "Oh, yes, I like that." Drunk on the freedom of choice, I went straight for blue jeans and tie-dye T-shirts. Ms. Nees subtly made a helpful suggestion by pointing out that I might want a dressier shirt and pants to go with all my casual clothes; that is, if I thought so, too. And then there was the jacket I selected, a cool-looking article of fashion made of a popular substance of the day that we called Pleather, a vinyl that appeared leatherlike.

My only regret of the entire venture was thinking, What

would Freda and Mona, Jessie, Fat Kenny, Sonya, William, and Michael say?—knowing, of course, that my friends would never be able to see me dressed so handsomely.

The saleslady made a memorable comment when she rang up my purchases. "Goodness," she said with a smile to Ms. Nees, "we know what his favorite color is. Blue!"

Me, a favorite color? Sure enough, blue was the dominant color of my new wardrobe. Little did I know that I would be officially restricted to that color throughout many years soon to come.

For all of this, I would be forever grateful to Patricia Nees. From the time she met me on that fateful day in March, she followed my case closely, registering her ongoing concerns:

> Antwone tends to fluctuate between feeling self-assured and confident about his abilities, potentials and expressions of self-doubt, uncertainty, and feelings of worthlessness. It is speculated, based upon test behavior, that Antwone is a "walking pressure cooker." . . . It is strongly suggested that formal therapy be initiated right away to help deter his exploding in an unprotected environment.

Ms. Nees may have sensed all of this early on, even in our first conversation together, when I sat beside her that morning with the drone of the rain in my ears. After we talked about me, from my perspective, and about some of the possibilities for my placement, she began to read from my file, telling me things about myself that I didn't know.

"Your middle name is Quenton," said Ms. Nees, approvingly.

Quenton. I loved the sound of it. Until now, I didn't have a middle name. I could remember a morning in Mrs. Profit's class when she read out middle names and some of the kids didn't know that they had them; I wished, at that time, that I had one. Quenton. It was different, rich and royal sounding, no average middle name. If only my friends knew that I had Quenton.

Ms. Nees flipped to another place in my file. My mother's name, she told me, was Eva Mae Fisher, and she was incarcerated at the time of my birth. In 1959, just after I was born, my mother had told her social worker that she named me Antwone after the real name of Fats Domino, because that was her favorite blues singer. Quenton was for a boy in her elementary class whose name she liked.

Silent and thoughtful, I looked off through the window, where I saw the rain was falling even harder.

Patricia Nees watched me as she offered up pieces of myself, pieces she somehow knew had been withheld all these years. Maybe she could hear my soul crying for connection; maybe she knew that after having been cut loose from the only family I had known, what I needed desperately were threads that connected me to my real family. Maybe she sensed the one question my entire life had deprived me from answering: Who am I?

And it was in this context that I came to know my father's name, Edward Elkins. In an interview with her social worker subsequent to my birth, my mother claimed that Edward Elkins was my father but that he had been shot and killed two months prior to my birth. Eva Fisher told her social worker that news of the murder had been published in the *Cleveland Call and Post,* a black newspaper, on June 13, 1959.

I received this information at the age of sixteen as if being handed an unwanted baby, me, not sure what to do with him, but unwilling to let him go unattended. In the meantime, I committed to memory every date, word, and syllable of what Ms. Nees was saying to me. It may have been my intense look of concentration or maybe a faraway glint in my eye that caused her to put down the file and ask me, "What are you thinking?"

"Seems like it was raining all the time back then," I said.

"It wasn't," she said, even as she nodded with understanding. Then she added, "You probably feel that way because it's raining today."

WHO WILL CRY?

Who will cry for the little boy?
Lost and all alone.
Who will cry for the little boy?
Abandoned without his own?

Who will cry for the little boy?
He cried himself to sleep.
Who will cry for the little boy?
He never had for keeps.

Who will cry for the little boy?
He walked the burning sand.
Who will cry for the little boy?
The boy inside the man.

Who will cry for the little boy?
Who knows well hurt and pain.
Who will cry for the little boy?
He died again and again.

Who will cry for the little boy?
A good boy he tried to be.
Who will cry for the little boy.
Who cries inside of me?

THE RAIN THAT FALLS

five

The first thing you notice when you're homeless is how long the nights are. It's hard to realize that a night can be so long; but in time you get used to it. You don't really sleep, especially in the beginning, because you wake up every fifteen minutes worried someone will come upon you. Your imagination runs wild with what terrible things would happen if you fell asleep and let that happen.

The world at night when you're without shelter feels like the Twilight Zone, another dimension, another planet, where the normal laws of time and space don't apply. When the sun comes up, you're so happy knowing that soon people are going to be out and you'll be back in the world again, on terra firma, although you're tired and worried about how fast night comes again.

Before you know it, you're seeing the shops close down, lights diminishing down streets, cars becoming fewer and fewer; on residential blocks you enviously watch working mothers and fathers pulling into driveways, arriving home to their families; and you stand outside talking to them in your head, saying, Don't go inside, not yet, stay out a little longer!

Your hearing changes as the general noise of the workaday world goes silent and other sounds become more pronounced. A car engine sputtering. Tires squealing around turns. Even sounds that are far away: distant trains, speeding cars, gunshots, police and ambulance sirens.

Soon you get used to the night smells. You may notice that broken glass has its own smell. The various smells of wine bottles—Thunderbird, Night Train, Boone's Farm—mixed with the other varieties of liquor and beer are distinct, too. These smells compete with the smells of rats, wet plaster, and rotting wood, the smell of a hollow place. And you can hardly avoid the recurring smell of human feces and urine.

Another thing about homelessness is that you lose track of what day it is. Without structure, Wednesdays feel like Saturdays, one week is no different from the next. Events of yesterday blur, often out of sequence, and no matter how you look at it, the day is never long enough because, when you're by yourself, your main preoccupation is night coming.

Night holds a separate world that can be far more brutal than sleeping in abandoned storefronts and alleyways and on park benches. It is the criminal business world where hardened humans prey upon the weaknesses and misfortunes of others, a world of hustlers, pimps, prostitutes, junkies, and murderers; a world populated by men and women, many of them young, who lost their way a long time ago. The night is host to this soulless world where parents loan their children to deviants for drugs or whatever else satisfies their empty hearts. This is a world you may never want to know, but it's a world that exists; it exists everywhere and probably has always existed.

When you're homeless and you're a kid, that world is waiting for you and is always on the lookout for new recruits. If you're a girl, God have mercy on you. If you're a boy, God have mercy on you, too. Depending on what kind of boy you are, you might survive; but if you're a girl, probably not. You don't need an invitation to come in, doesn't matter what you look like: fat, small, black, white, tall, Chinese . . . if you can breathe, if you're young and homeless, you're drafted.

Usually, you cross the threshold unaware and you're there, already a part of it. It's ready for you with a well-worn training program that dictates what you're going to do, where and when you're going to do it, and what will happen if you don't. So you do it—if you don't want to be homeless and lonely, if you want to eat, and if you want to avoid the seemingly endless nights and the smells and sounds they bring.

In my opinion, homelessness is preferable to being sucked into the machinery of the night. As a matter of fact, I think everyone should experience being homeless and going without. You get a different perspective on everything and a different appreciation for everything. You come to understand that you can be living in a house and still be homeless, as I was in the Picketts' home and in the institutions where I later went to live.

You learn what it feels like to be invisible to all those carefree or self-preoccupied people walking by you, maybe sidestepping the spot where you stand in order not to see you, or driving obliviously by you. Depending on your state of mind, you might prefer being invisible—as I did. If you've never been on the street, perhaps you may not see its world and all its inhabitants. But once you've been there yourself, you develop an extraterrestrial vision, and you see everyone who lives there. You retain your otherworldly eyes, even after you leave, and no matter how your own circumstances improve, you will continue to see the world that lies in shadows just beyond the gates; and you understand how easy it is to be pushed back out.

With the criminal underworld so ready to prey on unprotected minors, offering money and housing in return for work, the sight of a homeless seventeen-year-old may not be so common. As it was, most people I encountered didn't know I was homeless. Even the wino who slept for a while in the same storefront I did was sure I was a runaway. He kept trying to convince me to go back, no matter how bad it was at home. It took me a long time to convince him that I didn't have that option.

Indeed, given my age, the narrowing of my options had been the prevailing theme of my case from the moment Patricia Nees first raised the issue of finding placement for me. This was the reality that she and my previous social workers understood better than I did; it was the reason I had not been removed earlier when suspicions were raised about Mizz Pickett. At the same time, I got the impression that once I was away from the Picketts, recounting my own version of at least some of the events, Ms. Nees and her supervisor were alarmed that action hadn't been taken sooner, when there might have been more options.

Their great concern, I came to understand, was that I be given a voice in determining my future. It was touching and liberating on one hand, but bewildering on the other. "So where would *you* like to go, Antwone?" Ms. Nees asked over lunch at a diner not far from Metzenbaum Children's Center, the orphanage where I knew my placement to be only temporary.

She asked it as if I had the answer, as if I knew the option she didn't.

"I don't know," I replied, assuring her that I didn't have the answer, to which I quickly added, "You can send me to the moon, I just don't want to be around here any-more."

If there had been a doubt in my mind about staying in Cleveland, God cleared it up that day I visited Boys Town, after Mizz Pickett and her smirk materialized in the parking lot. The message was emphatic: Not only did my survival depend on my getting as far away as possi-ble, but, more to the point, it depended on my refusal to accept that she had been right about me. To prove her wrong, I had to go somewhere else, a place where I could reinvent myself and start again.

During the four months that Ms. Nees undertook the challenging process of finding such a place, I stayed at the orphanage, a way station both literally and figura-tively. To a large degree I was dead. Spent. Wrecked. Traces of the hopeful me shrank deep and small within myself, hiding from layers of hurt and anger about every-thing that had happened, as I spun my own cocoon from the threads of things I imagined for myself that would prove Mizz Pickett wrong. Without knowing I had begun my own transformation, that one desire became my rea-son to live. But I was so fragile that just thinking about her upset me to distraction, throwing me instantly into ab-ject depression.

That was the state of mind in which I found myself an unenthusiastic participant in a baseball game one Satur-day in the large field behind Metzenbaum Children's Center, where volunteers took turns coaching various sports and activities. Standing in the outfield, lost in thought, I felt little connection to my fellow orphans, who seemed to be enjoying themselves on the warm spring af-ternoon. They were all younger than me by at least a year, something I knew from the fact that I was the only kid who had to leave the center for classes at nearby East

Tech High School; the other teenagers attended school on the premises.

According to Mrs. Brown, the kindly woman who worked the evening shift as a sort of dorm mother, I was much more mature than the other children. Because I never said much, what she probably meant was that I wasn't silly or wild like a lot of the kids.

"Hey, you!" shouted the guy coaching us, waving his arms in the shape of an X to get my attention. The ball was flying toward me, and the batter was approaching first base. I walked halfheartedly in the direction the ball was headed, allowed it to land and bounce several feet, then picked it up and tossed it to the pitcher.

"Look alive!" the coach yelled in exasperation. "You gotta run for the ball!"

The next time the ball was coming toward me, I tried to pick up my step and show some vigor like the other kids, but I felt so drained, so tapped out, and my thoughts dragged me down further.

Finally, after more hand signals and yelling, the coach jogged out to me, saying briskly, "Quit acting so cool. When the ball comes your way, go for it, jump, run. Hey, you're not that cool."

Cool? I thought. Who gave a shit about cool? I didn't even want to play.

One of the few things I cared to do in my spare time at Metzenbaum Children's Center was to help the custodian with his cleaning. Physical labor relaxed me and gave me a space to be in my thoughts, with no one asking me questions I couldn't answer. In the evenings, after the chores were done, one of the other things that offered me some comfort was going to sit by Mrs. Brown in the communal room, a circle to which the different sleeping and eating corridors led.

236

A gentle-mannered, brown-skinned woman, Mrs. Brown used to knit while she was on her shift and often talked about her son, of whom she was fiercely proud, and who played in the NFL. Our conversations were one-sided: She talked, I listened. I probably had said no more than a sentence to her over several weeks and then one night, a stew of feelings bubbled up from inside me and burst forth into words that I had to speak.

Where it came from was complex. First there was the contrast of Mrs. Brown, a motherly woman quietly knitting and talking about her dearest son, mixed with memories of Mizz Pickett's scoff that I would never amount to anything; then there was a train of thought new to me, pictures from my imagination that showed me rising to a place of importance and success. Improbable as it was, I had to make myself believe that it wasn't impossible.

In the past my daydreams had been flights of fantasy; these new imaginings were serious, real, whatever I could conjure to make things different from that which Mizz Pickett had promised. I saw myself becoming a family man, a strong, loving husband and father, in a secure, love-filled home, as I—a man of stature, substance—provided for all my loved ones' needs. I saw myself popular with friends and neighbors, a man of the community, a contributing citizen of the world.

I didn't believe it; I couldn't. But I wanted so much to believe that I looked up at Mrs. Brown and announced with an intensity that appeared to surprise her, "You're gonna read about me one day."

Mrs. Brown paused in her knitting, not sure how to react. "Well," she said, a slight question in her tone, "I hope it's for something good."

"Yeah," I said, with more vigor than I'd felt bursting out of me in a long time. "Yeah," I repeated, "for something good!" In saying it to her, it was as though I had a butterfly in my hand and my words were that butterfly that I was freeing from myself. I wanted to tell her, *Look at what I have, it's beautiful! Of course you'll read about me for something good.*

Mrs. Brown thought about it, resumed her knitting, and nodded dreamily to herself, as if she was picturing the day she'd read about me for something good.

It was almost miraculous. She believed me. She believed, even when I didn't. And so it was, in that time and place, without knowing it, I stumbled onto a way to trick myself out of despair. Like that person who lies so much about something he eventually doesn't know the truth of it anymore, I saw that if I could learn to convince others that my future wasn't hopeless, in time I could convince myself.

To do that I would have to overcome not only all of Mizz Pickett's negative messages but also, I learned at the orphanage, the prejudices of the white world that, until now, had never affected me.

My first encounter with that reality came as the result of a dispute with one of the boys in my wing, who happened to be white, about cleaning the bathroom. That was the regular chore he and I had been assigned to do on alternate days. Even though I had cleaned the bathroom to a spotless shine the day before, on the day that it was his turn to clean, he decided he wasn't going to do it.

It was getting late and the lady who was sitting with us that night, who happened to be white, reminded him that he needed to do his chore.

With a jut of his chin over at me, he replied, "Why don't we make Antwone do it? That's why we brought those niggers over here."

All those years of Mizz Pickett demeaning me had in

no way prepared me to hear that remark from somebody of a different race. Furious, I could feel my ears burst on fire as I ran after him, determined to kick his ass. About two years my junior, the wiry kid saw me coming and began to run. I gave chase, caught up, and threw him down to the ground, at which point the lady intervened. Although she did make him clean the bathroom, she didn't reproach him for what he'd said. So the following morning I went after him again and gave him a few good knocks, but not to my satisfaction.

That same morning I was called into the office of the center's director, Ms. Day. Standing in the office when I arrived was the lady who had been on duty the previous evening. Ms. Day gave me the opportunity to speak first, asking, "Can you tell me what happened last night?"

"Yes," I said, grateful to be heard, and I told her what the kid had said. Ms. Day's face became very stern. With a glare, she turned to the lady standing there and asked if it was true.

The lady looked at me, then at Ms. Day, and told her, simply, "No."

Though my head was down in my usual stance, I dared to lift my eyes toward her in disbelief.

Apparently troubled but still unwilling to take the lady to task, Ms. Day opted to remind me that fighting was not tolerated at the center. Then I was dismissed.

I'd heard plenty of stories about racists and bigots, especially about the South, but this incident was my first time to experience the white world treating me unfairly because of my color. The whole thing reeked to me of terrible injustice, adding a new brand of distrust to the way I looked at others.

In her report from this time period, Patricia Nees noted that although I was always so guarded about my feelings,

I let her know wherever she sent me had to have some black people there, too.

She soon located a reform school that was two and a half hours away, just across the Ohio border in western Pennsylvania. The distance was a plus for me, as was the fact that there was a healthy racial mix of boys and staff. But Ms. Nees had misgivings about sending me to a reform school where I would basically be the only resident among hundreds who hadn't committed some kind of a juvenile offense. In the record, she also acknowledged her opposition to my placement in a large, impersonal institution but understood, she said, when reminded that resources were so limited.

When they presented the idea to me, I didn't really care that the other kids were there because they'd gotten into trouble. Then Ms. Nees took me to see what I thought up close, a trip she described in her report:

> On 5/5/76, Antwone was taken to Grove City, Pennsylvania, for a pre-placement visit at George Junior Republic, a private male institution offering complete child care facilities. Antwone seemed responsive to the idea. He accepted, as he does everything, with a fatalistic attitude.

While my attitude may have been fatalistic, I loved that it was so far away from Mizz Pickett and I was genuinely impressed with the physical setting. The place was in a rural area on a huge parcel of land, with trees, fields, streams, even a farm. There were big self-contained dorm buildings, each at some distance from the other, along with an infirmary at one location, plus a vocational school that was separate from the high school. There were nonuniformed guards, of course, who patrolled at all times and drove the perimeter in institutional cars. Even so, seeing that the kids were given freedom to move about within the confines of the property, I preferred the environment because it was a major contrast to city living.

That May visit to George Junior Republic turned out to be the last time I saw Patricia Nees, since placement there necessitated that I be transferred from the Child Welfare division of social services to the Institutions Department and that a new social worker be assigned to my case. None of these technical details were explained to me. All I knew was to pack the suitcase they gave me at the orphanage with the clothes and belongings I had acquired in the preceding months and be prepared to leave on June 7, a date that stood out for me, as it happened to be Dwight's birthday and was a cause to think of him and wonder where he could possibly be.

Early on that overcast, muggy morning, while the other boys in the large room continued to sleep, I silently dressed and made the bed, meticulously as always, the sheets and blanket pulled tight as steel. After washing up, I sat myself at the edge of the bed, my suitcase on the floor by my feet, and waited for Ms. Nees, watching the rest of the boys rise and leave for breakfast.

At the appointed time, I was alone in the room waiting when, instead of Patricia Nees, a man appeared in the doorway. Except for the light brown color of his hair, this guy was a dead ringer for Columbo.

I'm thinking, Damn, when did Peter Falk become a social worker?

"Bill Ward," he says, by way of introduction, then gestures for me to follow him. We walk together in silence down the corridor to the main lobby, past Ms. Day's locked office, down the stairs to the landing, where we exit the building through the double doors and outside into the morning drizzle. Clearly a man of few words, in his rumpled attire—rumpled white shirt, rumpled gray suit, and tie—and his well-rained-on London Fog trench coat, Bill Ward also appears to be a man less concerned with fashion than with practicality.

Without ado, he opens the trunk of his car, the usual white four-door sedan provided by the state of Ohio to social workers, throws my suitcase in, closes the trunk, and gestures for me to get into the front passenger seat. At last, from behind the wheel as we drive down Euclid, me at his side, looking out the window at Cleveland in all its gloom, Bill Ward speaks.

"Hungry?" is what he says. I shake my head no. "Had breakfast?" he speaks again. I shake my head no again. To that, he shrugs, as if to say, Suit yourself.

We travel south and east, with the Glenville neighborhood fading farther and farther behind me, as some of the last familiar sights of downtown blur in my rainy periphery: Public Square, with its statues of Cleveland's founding fathers and the Old Stone Church; the concert halls and the museums I visited in the glory days of Mrs. Profit's class; Cleveland Stadium, where I have never been to either a Browns or an Indians game, but have always noticed the way INDIANS is emblazoned in swirly script, alongside neon beer logos and the face of the mascot, Chief Wahoo, with his huge teeth and grin; and the murky Cuyahoga River snaking under bridges, across our departing path through the city. By the time we veer onto the on-ramp for the interstate, my head is turned all the way back to look through the car's rear window at Cleveland's skyline. With the spires of Erie View Tower and Terminal Tower poking up into the mist, this last sight resembles, not by accident, Metropolis, home of Superman, who had been an orphan, too.

With a sharp, cutting pang I see on the canvas of my mind the faces of my friends: pretty Freda; her prettier sister, Mona; tough Jessie; easygoing Fat Kenny; his thoughtful sister, Sonya; my younger sidekick, Michael Shields. They come into my senses as a reminder that I never got to say good-bye, as though they're some kind of farewell committee, the only family I know. Already I miss them more than I can bear. And the thought that I'll never see any of them again is too excruciating to accept.

In the midst of my tidal wave of sadness, Bill Ward thrusts a no-nonsense, no-frills question at me: "What are your plans for the future?"

Plans for the future? I look at him like, Give me a break.

"You're going to be seventeen in two months," he states. "When you're eighteen, you'll be on your own. You need to have a plan."

My mind reels. I resent him, some TV-detective-looking guy, thinking he knows what I need to do. A part of me knows he's saying something important that I ought to hear; the other part can't even begin to conceive of a future beyond this moment. Hoping to get him off my back, I give him my best solemn nod of agreement.

He ain't buying. Though he's brusque and to the point in his manner, Bill Ward decides we're going to chat. "George Junior has a strong visiting program," he says, explaining that good behavior is rewarded with visiting passes to go home. Then he suggests I should think about whom I'd like to see on holidays and weekends.

Bothered by the whole topic, I turn to him and point out, "I don't have no home."

"What about foster brothers or sisters? You could get together on the holidays."

"No."

"We could look into locating your natural family, maybe find some relatives you could visit."

I look at him hard. Start to speak. Feel a choke in my voice as I say, "I don't have no family."

A brief silence ensues. The rain has stopped, giving way to a sunny summer day shining down on the flat farmland of eastern Ohio. I roll down my window, breathing in the clean air, fighting back my tears.

"Antwone," I hear Columbo say, and turn to him as he offers, in the kindest tone he can muster, "stop feeling sorry for yourself."

Sorry for myself? Hey, dude, I want to say, you don't even know me. How do you know what I'm feeling?

Loosening his rumpled tie, he gives me an understanding nod. Then he looks back at the road ahead of us and speaks to no one in particular, saying, "Yeah, don't feel sorry for yourself. It doesn't do any good."

I sit there mad for the rest of the drive, not yet willing to see what I would later—that God himself must have chosen Bill Ward as his special messenger to tell me what no one had ever told me before and, by giving me that vital piece of information, to help transport me safely into the unknown.

And so, for the next twelve months, George Junior Republic became a form of refuge, a place where I could rest and recuperate, unmolested—as the social workers might have said—my last home before homelessness.

My first few months at George Junior were spent trying to fend off a new panic about the future that Bill Ward seemed to be such an authority about. Later I could see that he didn't necessarily think I felt sorry for myself;

244

he was only warning me that, in case I did venture down the self-pity track, it would be an indulgence my situation couldn't afford. Oddly enough, given the coincidence of his last name, Bill Ward never used the term "ward of the state" in reference to my status, in order to specify that upon my eighteenth birthday the state of Ohio would no longer have any responsibility for me. That was still too sophisticated for me. But I was starting to perceive that eighteen was the mystical, magical age when something pivotal was to happen to me. And whatever it was, I understood, as Bill Ward had intended, that I had to begin planning for it right away.

At first, these new worries about the future only intensified my despondence about the past, causing Bob, the plainspoken black teacher who taught us food service at the vocational school's cafeteria, to confront me while I was in the middle of washing dishes one afternoon.

Bob approached me, leaned into my ear and got right to the point. "If you need to talk to somebody," he said, "you should just open your mouth and talk."

He waited. I kept washing the dishes, willing him to leave me alone. That was the place I was in, not wanting any attachment, not to social workers, not to teachers, not even to the other kids. And definitely not him.

Bob wouldn't leave me alone. "You can't be walking around here depressed," he said, "it just doesn't make any sense."

Depressed? Me? My initial reaction was, What does he know? Bob knew a lot about food service, instructing us in aspects of food preparation, kitchen maintenance, how to wait tables, even how to be a maître d'. But he didn't know nuthin' about me.

Bob said again, "You should talk to somebody."

Not looking up from the dishes, I spoke to him in my

head: Why are you in my space? It jarred me to hear that there was anything noticeable about me to reveal what I was feeling. Beyond that, I didn't want anyone trying to be my parent. I didn't need any authority figures. I'd just gotten rid of one, thank you very much. The way I figured, I was done with anybody else being in charge of me. In fact, I was done with adults altogether.

The irony was, Bob was right. I hadn't known what to call what I was feeling, but now I had a name for it. My guidance counselor, also named Bob, a bushy-mustached white college student with round wire-rim glasses who never smiled, whom I was assigned to see regularly, also commented that I seemed depressed. He did add that it was understandable. That relieved me enough to know that it wasn't going to be fatal, but not enough to open up and talk about it, in spite of his encouragement to do so.

By the end of the summer, after passing an uneventful seventeenth birthday, I was so low that not even those tricks of my imagination about eventually making a success of myself could ease my mind. It was at this time that a three-day visit to Cleveland was arranged for me at the home of Mrs. Ewart, a friendly, good-hearted black woman in her early forties who had been a volunteer at Metzenbaum Children's Center.

On the drive there, Bill Ward may have been hinting that Mrs. Ewart was thinking about taking me in as a foster son. If he was, I didn't know that; I didn't know that she was trying to see if we would be compatible. Mizz Pickett had done much too proficient a job in telling me that no one would ever want me. It turned out Mizz Pickett was wrong. After spending a quiet three days at Mrs. Ewart's comfortable, attractive home in one of the newer, nicer Cleveland neighborhoods—just me and her—she said to me, shortly before Bill Ward came to take me back to George Junior, "You have a home, if you want one." She looked at me thoughtfully, then added, "Here."

It was an ideal situation; she would have been exactly the kind of foster parent I needed. But even knowing how fortunate I was, a teenager, to be made such an offer, I was too emotionally damaged to accept it. Embarrassed, feeling totally unworthy, I said nothing.

Bill Ward brought it up again on the drive back, that she sincerely wanted me to come stay with her.

"As my foster mother?" I asked him, guardedly.

"Yes. What do you think, Antwone? You want to come back to Cleveland?"

"No," I answered, and left the matter there. I couldn't risk another foster home, no matter how nice the people seemed. This would mean having nowhere to go on holidays. But holidays had never meant anything to me before, why should they now? Other than missing my friends so terribly, there was nothing for me in Cleveland.

As summer turned to autumn, I cocooned further into myself, spending free hours on long walks through pastures and down country roads, petting stray goats, shooting baskets alone, drawing and doodling in my room. I drew portraits of Freda and the others from memory and designed intricate crossword puzzles made up of all their names. Since nobody else at George Junior knew the answers, I'd come back later and fill in the squares myself. On the basketball court or at the gym, I made up wishing games about whether I'd see them again.

There I am alone, a lonely seventeen-year-old kid, dressed in long-legged and long-sleeved sweats, dribbling down the gym court to the squeak of my sneakers on the wood, shooting a long shot from center court. As the ball rises in its upward arc, I think, If it goes in, I'll see Freda again. And when the ball misses, I try again with a hook shot: If it goes in, I'll see Freda again. Then I'm at the free-throw line. If I make it, I'll see Jessie again. I make it. Another free throw. If I make it, I'll see Michael Shields again. I make it. Now a layup. If it goes in, I'll see Fat Kenny again. When it misses, I try another layup, promising myself, If this one goes in, I'll see Fat Kenny again.

Even on days when I wasn't shooting well, I made up my own rules so that I was able to win, ultimately convincing myself that I would certainly see my friends again. Those games kept me alive, as did listening to the romantic songs I associated with my friends and our time together—songs like "Do It Baby," "Dream Merchant," and, above all, every Dramatics record I could get my hands on.

As much as I kept to myself, the other kids, teachers, and counselors at George Junior seemed to like me a lot. In the room I shared with three roommates, one black and two white, the four of us got on well. Ulysses, the black teenager, who was from a neighborhood in Cleveland a block away from Glenville, went so far as to write to his mother, asking if she would adopt me. When she wrote back, explaining to her son that she wasn't in a position to do that, he was much more disappointed than I was.

George Junior designated me an Honor Boy, meaning that because of my status, I could leave any time my classes weren't in session and go into town on my own. There were other privileges given to those who didn't earn any demerits. One such perk was being able to attend periodic dances held in Waynesburg at the girls' school, an institution similar to ours.

The first time I attended one of these dances, I spent a half hour battling my shyness. Finally, as naive and tongue-tied as I felt when it came to girls, I forced myself to ask an attractive, friendly girl to dance. The moment she nodded yes and I turned for her to follow me out to the dance floor, I realized it was a slow song. A slow dance? I'd barely ever danced in public before, let alone slow-danced. But at the same time, it occurred to me that nobody knew me here. Maybe some of them knew I was shy, but nobody knew how shy I really was. Actually, as I allowed the girl to lead me, following her steps closely, I saw that she considered herself lucky to be dancing with me. She even glanced smugly over to her girlfriends who were waiting to be asked to dance, tilting her head in my direction, like she wanted them to notice me—a tall, strong, dark, nice-looking, well-mannered guy.

Thus I found my chance to start again, to reinvent myself. And so, at that very dance, I went from being excruciatingly shy to being quietly reserved. A quantum leap. The only things that kept the experience from being perfect were the erection I got from having her tender body so close to mine and the long walk from the dance floor back to my spot on the sidelines, where I was sure everyone could see the hearty pool stick protruding from my left pants pocket.

Back at George Junior with the rest of the boys, I tried to adapt myself to their terms, pretending I was tough and cool, too, showing how I could curse and make dirty jokes like them. Then, becoming really uninhibited, I'd riff wildly with imitations and stories of Mizz Pickett.

They all knew I was a good kid, and I knew they were mostly badass kids. But just to prove I wasn't any kind of pansy, one night I decided to give it to this whiny punk kid who was getting on my nerves. Taking a wet towel, I popped it—*smack*—on his butt when he was lying on his bed.

Our dorms, which were known as cottages, were arranged with sleeping rooms, without doors, in the shape of a square. That way, the cottage uncles—more or less prison guards—could patrol the halls and see into the rooms at all times. Naturally, when I popped the whiner with the towel, I knew Uncle Dave, who was on duty that afternoon, was nowhere near.

Uncle Dave had already been a lesson to me about not judging a book by its cover. From his looks the first time I saw him—skin so white he glowed, and dressed like a cowboy—I guessed he was just a redneck. Plus, the dude listened to nothing but country music. After a while, however, I noticed that Dave's appearance and musical tastes had nothing to do with his fair, low-key, warm personality. Treating everybody with the same respect, he had a gentle manner and yet, without playing politics, he had power. For example, whenever there were infractions he wouldn't punish an individual, he'd punish the whole group.

When I snapped the scrawny kid with the wet towel, it couldn't have really hurt enough to get Uncle Dave's attention. But lo and behold, the kid screamed like he was having a baby. The next thing we knew, Dave was there, in his cowboy hat and boots, hauling the whole dorm out in the hallway as he walked up and down the line like a country-western drill sergeant. He made himself clear. Nobody was going to mess with the little guy. But instead of making the kid rat, he asked for the perpetrator to identify himself. If no one did within the next half hour, everyone in the dorm would be required to do the shine line, a fun name for the not-fun task of buffing the floor. Then we were dismissed. When these things had happened in the past, nobody ever fessed up and certainly nobody ever snitched on anybody else. But within five minutes of thinking about it, I decided to turn myself in.

Uncle Dave was cool and let me off without punishment. "All right, Fish," he said calmly, "but don't let there be a next time."

I got the feeling he understood I was just being a kid. On the other hand, I got the feeling that he was worried about my behavior being influenced by some of the more problematic kids and that he was keeping his eye on me.

A week later, some of the guys were complaining about all the rules and regulations at George Junior, all the classes they made us take, and not letting us have any fun. "Man, it's fucked up," snarled one kid.

I was about to join in when I changed my mind. The truth was that for a lot of kids, this place was a chance, maybe a last chance, to make something better of their lives before adulthood was upon them. The George Junior philosophy was: It's up to you. You got into trouble, and now you have to pay for that. So while you're here, we're going to show you that you have options, some of the choices life has to offer—whether in an academic way, in our classes at vocational school, like auto mechanics, welding, and food service, or in being able to enjoy the beautiful natural setting.

In the midst of their complaints, I said, "You might not understand, but I'm glad I've got this place, 'cause I don't have nowhere else to go."

Another kid said, "You like it here? I never thought I'd hear you say that, Fish."

"Well," I reminded them, "I don't have the choices you guys have."

They understood. After that, we sat in silence, some of them maybe even taking a different look at their own situations. My own words brought home to me the fact that, unlike them, if I made a mistake I didn't have anyone to come get me out of trouble. From then on, I began to set stringent rules for myself, many of which were to last a lifetime. In this unexpected way, George Junior gave me the chance to become a parent—to myself.

First I decided that I wasn't going to be out late at night. Nine or ten at the latest would be my bedtime. I wasn't going to smoke cigarettes, do drugs, or do anything to get arrested. The worst trouble I could imagine was getting a girl pregnant. Since I was the only person taking care of me, I certainly couldn't take care of kids. So sex was out of the question.

Perhaps it was no coincidence that during this period I received one of the only pieces of correspondence I had ever gotten, let alone at George Junior. The return address on the large brown envelope was that of the Child Welfare division of social services in Cleveland. Inside, along with a brief note from Patricia Nees wishing me well, was a copy of my birth certificate.

Until now, I had never been in possession of a legal document. And this was so much more—it was proof that I, Antwone Quenton Fisher, existed and had existed since the day on August 3, 1959, when I was born to a woman named Eva Mae Fisher. The delivery to me of my birth certificate from the state of Ohio also sent a loud, clear signal of the very thing Bill Ward had been trying to tell me—that my eighteenth birthday was approaching, fast. Now, with sudden urgency, I began to question how it would impact me. My main concern came from realizing that I was only in the eleventh grade and that in the summer I would turn eighteen without having finished the twelfth grade. Did that mean the state of Ohio would no longer pay for me to attend George Junior, even if I hadn't graduated? Could they turn me loose without a diploma?

I asked Uncle Dave, but he didn't know the answer. Neither did Bob, my guidance counselor. The next person

I approached was my history teacher. Dumpy and not very attractive, he appeared to me as probably the most knowledgeable person at George Junior. Or the world. In class, he could talk nonstop with intense seriousness about the subject that he found so fascinating. That was, until we discovered that if we enticed him to talk about his new wife, he would forget history and talk euphorically about her. Our history teacher looked like the kind of guy who had trouble finding someone to marry him. But, miraculously, he had found her. For us, it was much more interesting to hear about their courtship, wedding, and setting up housekeeping than about the battles of World War II.

Nonetheless, I suspected that his knowledge of history might give him insight into how I should plan for my future. But instead of asking him what he knew about the ins and outs of Child Welfare, I decided to present him a possible plan for myself. The idea came to me from reading *Jet* magazine, after noticing that many of the pictures were taken by the same photographer, that I could do something like that. After all, I was good at art and at arranging my physical environment. What really excited me was the notion that I could travel to faraway places, even foreign countries.

After school was over one day, I spotted my history teacher on his way to his car and ran after him. Catching my breath, I just blurted my feelings and questions out all at once, telling him, "I think I want to be a photographer when I leave here and I was wondering if I could do that and, I mean, I really want to make it in life, you know, and I want to know what I need to do to do that."

"Well," my history teacher began, thinking it over, "you have the right attitude. If you keep that up, you will make it."

"Really?"

"Sure, with that attitude. You bet. As long as you keep your grades up and work hard. Absolutely."

Wow. I was going to make it. I was going to be a world-famous photographer. This didn't answer my question about whether the state of Ohio was going to let me graduate from George Junior, but both my study habits and my grades saw swift improvement.

Although the obvious person to ask about my status would have been Bill Ward himself, I was afraid of what his answer was going to be. In our earlier conversations, he'd already suggested I plan on taking the GED to complete high school and then enlist in the army. Hell no, I thought. It reminded me of Vietnam, a place of horrors I grew up hearing about. Bill Ward was only trying to prepare me for the reality that if I turned eighteen and wasn't in a foster home or the army, there would be hell to pay. Unfortunately, I couldn't see that yet. Fortunately, I was at least smart enough to know that I had to get my diploma before my eighteenth birthday.

Taking the issue into my own hands, as I had never done before, I arranged to speak with the principal of George Junior, a rugged ex–army officer by the name of George Tucci. As a matter of fact, Tucci looked like Sergeant Carter from the TV show *Gomer Pyle, U.S.M.C.*— high-and-tight military haircut, pits in his skin, a bulldog expression and way of talking.

After hearing of my dreams to become a photographer and my concerns about not being able to receive my diploma on time, Principal Tucci agreed to take the matter up with his vice principal and with social services.

He called me into his office for a meeting a week later. Sitting there at his side was Vice Principal Steve Slencak, a man who bore an uncanny resemblance to George Kennedy, the antagonist in the movie *Cool Hand Luke*. When I closed the door behind me, I saw Columbo was there, too. The three had come up with an option for me. I was to be given a series of tests. If I passed all of them, I would be allowed to graduate the following spring with the seniors.

Each of these three men gave me the same look. You can do it, their faces told me. I breathed hard, thanked them, and left the room. It wasn't just that I could do it. I had to do it.

With heightened nerves, I proceeded to take the tests over the next few weeks. Vice Principal Slencak had the honor of informing me each time whether or not I had passed. Every other day, I would be in the middle of some class and the door would open and his head would lean into the room. "Fisher?" he'd say as I jumped to attention. "You passed." Math, English, social studies, history, science, I passed each test. By Christmastime, I had succeeded in earning the right to graduate with the seniors.

During this period Bill Ward made his first and only entry to appear in my Ohio state social services file. It also turned out to be the last entry of the file:

Planning is for Antwone to remain at George Junior Republic until his graduation from high school which should occur in June, 1977. At that point he will enter the United States Military Reserve for a period of 6 months so that he may have the financial means to return to Cleveland to take care of himself, and enroll at Cooper Art School to reach his goal of becoming a commercial artist for which he does have the talent.

As planned, my high school graduation took place on June 7, 1977—Dwight's birthday again. It came much too soon. The state of Ohio paid for me to buy myself a suit from a local department store. It was a three-piece beige suit, over which I donned my cap and gown—blue and gold, the colors of George Junior Republic. The ceremony was held outdoors with many guests that included family and friends of the other students who were graduating. The highlight of the event was having my picture taken with Principal Tucci, a photograph I would long cherish, together with my diploma and my birth certificate—two documents I would soon decide to keep on my person at all times.

Partly excited to have arrived at this momentous occasion, I was too anxious to relish the experience. The next day, I knew, Bill Ward was coming to pick me up, not to take me to the Military Reserve—which I had resisted adamantly—but to return me to Cleveland, where I would begin life as an emancipated minor in a YMCA men's shelter. Since my eighteenth birthday was two months away, I had only that much time to find my bearings in the adult world. As the next few days would reveal, I would be forced into the adult world much sooner.

A lot of things have changed since you've been gone," Bill Ward says, by way of making conversation, on the drive back to Cleveland.

How much can change in a year? I think. But to be polite, I say, "Really?"

"Yep. For one thing," he observes, as we pull off the highway into a filling station, where a line of cars means we'll have to wait our turn, "you have to pump your own gas now." Shaking his head sadly, he tells me that atten-

256

dants no longer pump gas, wash your windows, or check your oil and tires. Now they only take your money.

Bill Ward hasn't changed much over the year. He's still in a rumpled suit and tie, still got his world-weary London Fog raincoat folded in the backseat, always at the ready, even on this warm, sunny summer day.

But as we approach the city limits of Cleveland, I start to see that things don't seem the same. It surprises me how much my year's separation from here has changed it for me. The cars are different. They're smaller and more compact, including a cool-looking Cadillac, the most streamlined I've seen. Even familiar sights appear to have changed. The older buildings look less historic and more run-down; the newer buildings look less like Superman's Metropolis and more like Anywhere urban sprawl.

Somehow these subtle differences trigger a wave of relief, which I take to mean that the Picketts are definitely gone. We're nowhere near Glenville, but I'm certain of it. Like a refugee who has fled his homeland, I rejoice in the knowing that it's okay to come out of hiding now and return. For the first time, I begin to contemplate actually showing up in the old neighborhood.

Just as my spirits start to rise, they evaporate the instant we reach our destination on Carnegie Boulevard, at the threshold of downtown, far away from the address on the same street at 105 where I went to see Dr. Fisher so long ago. Bill Ward parks the car across the street from an older four-story building bearing a small, vertical, weathered sign that reads, YMCA. Carrying my suitcase for me, he escorts me across the street and up the steps to the entrance. There he hands me several coupons, telling me how to find the diner where I can exchange them for meals.

"Better start looking for a job," he says, as a final reminder that my birthday is around the corner. "Call me if you need anything," he offers with a wave, and then scuffs down the steps, across the street, and into his car. Watching him drive away, I don't yet know that I have seen the last of Bill Ward, the last of the almost two dozen social workers who have had a hand in my rearing by the state of Ohio.

It's already starting to swelter outside. Inside it's hotter. Stifling. And it smells musty, a mix of sour body odor and pine cleaner. At the front desk, a sweaty, emotionless middle-aged man checks my name off a ledger, hands me a key, and directs me to a room on the third floor. On my way to the stairs, I walk through the lobby, where several of the shelter's residents are gathered together in a shabby, unsavory array, watching television. Cringing inside as I feel their eyes upon me, I hurry along, up the two flights and down the stained carpet of the corridor to my room.

As small as a jail cell, the room is hot, with a window painted shut. Here, too, the carpet and walls are stained and the musty smell is worse than downstairs. There is a bed with yellowing sheets and an old dresser. For the next hour, I put my things away, thinking about what I can do to improve the decor. Though I flash on my earlier thoughts about going back to the old neighborhood and looking for my friends, it escapes my grasp, like a speeding train I can't get to the right platform to board. First things first, I decide. First, I have to get myself together. Find a job. Earn some money. Go back in style. But before I do any of that, I have to walk through that lobby again and out to the diner, a decision I make only after much self-coercion.

At the diner, it's of little consolation that most of the others in this place look worse off than me, although they're not as scary as the group I encounter back in the lobby watching the movie *Logan's Run* on TV. Head down, eyes averted, I find a chair off to the side, where I can watch without being noticed.

Whatever's happened to these men, their goodness has plainly left them years before. They're wasted, with all interest gone of doing anything to rise above where they have fallen. Most of them are in their thirties or older; there are a few younger ones, but they seem just as lost as the older guys. One guy in his late twenties or early thirties, tall, light-skinned, handsome, looks out of place, especially in his expensive long leather coat, which is also out of season. Not sure what he's talking about, I notice him joking with a couple of the other men, but they're not laughing. In fact, they seem intimidated by him. As unpleasant as it is in my room, I opt to go up there and lock the door behind me, falling soon into an uneasy sleep.

I wake in the middle of the night. Got to pee. Dressed in slippers and my childlike pajamas, the blue-and-white patterned cotton ones I picked out a year earlier, I weave my way, half asleep, down to the bathroom at the end of the hall. There is a row of stalls and a row of urinals. Going for the nearest and most accessible, I pee abundantly in one of the urinals and am just buttoning up the fly of my pajama pants when a man's booming voice sounds from behind me, "Don't put it away."

I turn quickly to see a big, beefy black man who's been in one of the stalls the whole time. He's standing in his dingy drawers, underwear being the attire of choice at this men's shelter. I haven't yet picked up on the fact that pajamas and slippers are foreign here.

The man moves closer. "*Umm,* you *fiiine,*" he says, reaching for me as I turn around. He grabs me around my waist, pulling me toward him.

"What you doin'?" I say, pushing him away. "Get your hands off me!"

He sidles back a step or two, assuring me, "I'm not gonna force myself on you."

Force himself on me? Never heard that before. The implications pound through my brain as it hits me. "I ain't no fag!" I warn him, my fists clenched as I push past him, out the bathroom door, down the hall, and into my room.

Now, in my comings and goings over the next two days, I avoid the bathroom alone at night, even when I have to go. On my third night at the shelter, I see the same man and a shorter, thinner cohort in the TV area as I'm passing through.

He points to me, says something to the other guy, and they laugh. Then the other guy blows me a kiss. I continue on quickly to my room. I make sure the door is locked, pace the room to let off steam before dressing for bed and trying to force myself into sleep. Close to midnight, I'm still awake and have got to go to the bathroom so bad it hurts. The hall is empty as I start down it toward the bathroom. But halfway there, the big man and his sidekick appear from around a corner.

They start in together, whistling and razzing. "Hey, baby," the shorter man says, talking to me like I'm a girl, "don't run off, we jes' wanna talk to you."

"Now, I tole you," says the big one, coming toward me, "we don't wanna force ourselves on you."

The other one repeats him, coming toward me, too, "Nah, we ain't gonna do dat."

Force themselves on me? The insistence of repeating that line makes me sure the opposite is their intent. I've been through enough to know what Shakespeare meant by "the lady doth protest too much, methinks." Trying to appear unafraid, I ignore the two and continue toward the bathroom. The smaller guy blocks my path. The big guy looms behind me. Me, the sugar-coated romantic, I'm Smokey Robinson in my mind; girls are everything, the reason to live. Why anybody would want a boy, I don't know.

The shorter guy says, "Relax, honey, this ain't gone hurt. We jus' try'na get to know ya. . . ."

"I ain't try'na know you!" I yell, as they both lunge for me at the same time. Punching and kicking, I wrestle free and jet to the stairs and down them, taking the steps in twos and threes. I can hear the two of them thundering down the stairs right behind me. Down the hall on the ground floor, I spot the rear exit door and push the panic bar as I fly outside into the alley. The door closes abruptly behind me just as I realize it locks automatically. I'm in my pj's, alone in a dark alley, after midnight. The two men are nowhere to be seen. The front lobby door is locked at 10 P.M. Absolutely no admittance. Trying to keep hidden, I stay in the alley for the remainder of the night.

Sleep is impossible. As the hours pass, I thrash through my thoughts, rummaging in them like garbage, searching for some kind of solution. By dawn, I've hit on my only resort—to sneak back in and up to my room, get my stuff, and leave the shelter.

This is how I spend my first night on the street. It will not be my last, now that I have passed Homelessness 101 and learned the first lesson about why you become homeless. Not necessarily because of hunger or lack of shelter, but because of fear; because sometimes outside is less terrifying than inside.

six

In the span of this first long night in the alley I give up my other priorities, like finding a job to save money for art school and reuniting with my friends; instead, my focus now is on finding protection.

I find that protection in the lobby this same morning as I'm trying to sneak back in. Rather, the protection finds me. Going past the TV room to the stairs, I hear laughter coming from the group that's gathered there. When I realize I'm the cause of the uproarious laughter, I'm mortified.

"Where you been?" asks the tall, light-skinned man with the long leather coat. "A motherfuckin' pajama party?"

"A pajama party," laughs another guy. "That shit's funny, Butch."

The tall dude, Butch, ignores the others and turns serious, asking me why I'm dressed as I am, coming in from outside.

I tell him about the two guys and about spending the night on the street.

"All right," Butch says, looking more sincere. "What are you, fifteen?"

"Seventeen and a half."

"What you doing here?"

Glad to have someone be concerned, I tell him about my situation. Butch nods sympathetically as he hears that I don't have any parents and no one to look out for me. "Listen, little brother," he promises me with a grin, "don't worry about that. And I'll take care of them two niggas."

True to his word, those two will never be seen or heard from again, at least not around there.

Seeing Butch the first time, I assumed he lived at the shelter. But this morning it becomes apparent that he's only here to do business. Cool, smart, and articulate in his own way, he gets right to business with me, saying, "You gonna need to fix your situation up . . ." Butch pauses. "What's your name?"

"Antwone," I say, "but people call me Fish."

"Fish? I like that. Fish, you gonna need to fix your situation up. You gonna need some money. You gonna need a job. Now, it so happens I'm in a position to offer you an opportunity to work. You're a minor, that's good, you can't get in no trouble. As long as you keep your mouth shut, everything'll be fine. Anyone asks you anything, you don't know nuthin'."

Butch pats me on the shoulder, assuring me everything's going to be easy street from now on. Then he suggests I go wash up and change my clothes and he'll put me to work right away.

O ver the next week, I became acquainted with some of the activities and some of the other participants that comprised Butch's criminal enterprise. He ran prostitutes, numbers, drugs, stolen goods, and whatever else could be bought and sold outside the arm of the law. As his young protégé, I started out as a runner picking up cash from his hookers, as Butch called them. About seven women in all, they ranged in size, age, and color—mostly white, some black, some Hispanic.

In reality, Butch didn't give a shit about anyone other than himself. Blind to that at first, I was too caught up in the ambience of having him look out for me. Knowing prostitution was illegal, it worried me a lot that I could get into trouble. But Butch was big on the fact that I was a minor, repeatedly explaining if I got picked up by the police, nobody could do anything as long as I didn't say a word.

It didn't take long to see that the prostitutes were scared of Butch. But I was scared of them, mainly that they were going to try to have sex with me. Even looking at them made me nervous, with their shapely bodies, big boobs spilling over their plunging necklines, thick legs in their hot pants and miniskirts. They were real grown-up looking and their business was sex. Whenever they handed me their money, I stared at the ground, avoiding at all costs any innuendo that hinted they were making a pass at me. They didn't have to hint; to my chagrin, a few actually came out and offered to pop my cherry for me. One of the younger hookers, Rhonda, a beautiful girl in spite of the hardness in her eyes, tried to make friendly conversation that had nothing to do with sex. But that, too, seemed threatening.

After a day of picking up money, I had as much as eight hundred dollars to turn over to Butch. He'd give me twenty dollars in return. More money than I'd ever received in one shot. By the end of a week, I had a hundred bucks. More money than I ever could have been making at a legitimate job. So I stayed at the men's shelter, keeping my earnings in my shoe, and my birth certificate and diploma in my pocket whenever I left. With the money I was earning, I wasn't going hungry, even though, without supervision, my diet wasn't the healthiest: Ding-Dongs; grape and orange Nehi sodas; and Buckeye potato chips, a local favorite, named for the state tree, with a superhot barbecue flavoring, so fiery to the taste that only the sweet fruity Nehi could cool my palate after eating the chips.

Butch had a guy working for him whom we called Tank—a big gorilla whose real name I had the feeling nobody knew. Tank would school me every chance he got, warning, "You don't wanna be no player! Now whatchoo wanna be is a doggone pimp! See what I'm sayin'. Players get played and pimps, huh, dey get paid up in here. Hear me?"

He seemed to like me, the way he was giving me this personalized instruction. Since Tank was Butch's field marshal, it occurred to me that Butch was someone's field marshal as well.

Within the month, Butch also put me to work rolling joints. Along with his other young apprentice, I'd sit for long afternoons, sometimes competing to see who could roll the most, for a reward of fifty dollars. The other kid always won and his looked as professional as cigarettes; mine were just twisted up at the ends, neat but half-hearted. This got old fast. Licking the Zig-Zag papers, I grew to despise the taste of reefer and being in the smell all day. Like working in a sweatshop. But I was getting paid.

Then, toward the end of July, Butch gave me a new responsibility to add to the others. This involved selling playing cards with local addresses written on them. I didn't understand what these cards were for, but whatever it was, Butch had a steady flow of customers paying three hundred dollars apiece. The clientele was mostly white, middle and upper class. Some of them were regulars and Butch had me make sure I kept extra cards on hand for them.

That raised a problem for me one night when a new customer, a middle-aged white man, approached me on the corner near Quincy, saying that Butch had sent him.

"How many cards do you have?" he asked.

"Six, but I can only sell you four. Gotta keep two for regulars."

"Butch told me I could have six. Look, here's the eighteen hundred bucks." He started to shove the money at me. Suddenly looking nervous, he agreed to take the four, giving me twelve hundred instead.

That night when the regulars didn't show, I figured I'd have extra for them the next night. But the following day when I went to turn the money over to Butch at the usual spot on Quincy, where I found him with his entourage of subcontractors—his dope dealers, prostitutes, and the guys who acted as his field marshals—he went ballistic. Shouting and cursing me for being so stupid, he announced that he was going to give me a business lesson and proceeded to beat the shit out of me.

Butch's first punch sent me flying backward onto the sidewalk. I tried to get up but he kicked me back down, stomping me for several moments before straddling my body, punching me over and over with both fists on my ears, head, and chest. The whole time he was kicking and hitting me, the rest of the group, men and women who'd been acting like my pals all this time, stood by doing nothing, blank expressions on their faces, until at last, dazed, my vision blurring, I could no longer see them. Butch gave me one last wallop that threw me down into the street. Then he took off and the rest of the crowd disappeared. Everyone except Rhonda. She came over, helped me up, and examined the damage. I wasn't bleeding too badly from scrapes, but my eyes were swelling up, she said. From the pain in my side, I was sure a couple of ribs were broken.

Rhonda took a look around the street, making sure no one was in earshot, and then warned me to get the hell away from Butch. "You don't belong in the game. My advice to you is to get out now, unless you wanna wind up like Butch, in prison, or dead."

She was right, I knew. But in that numb, shell-shocked state, I had to ask what I had done that was so bad. "Why'd he get so mad about just two cards?" I asked her.

That was when I found out what the cards were. Rhonda explained. Some junkies who couldn't afford to pay Butch cash for heroin traded their kids out for a night or two, so Butch put their addresses on the cards and that's where the kids were. Customers who bought the cards used the kids for sex, pornography, whatever they wanted.

The beating I'd just gotten from Butch was nothing compared to what I felt hearing what those cards meant. I felt like dying. For what happened to me with Willenda; what happened to Keith; for all the kids whose names and faces I'd never know that I'd unknowingly helped Butch use.

Whatever innocence I'd salvaged through this summer died that day in that spot as Rhonda, having said enough, disappeared around a corner. That night, after taking with me from my room at the shelter only what I could carry on my body, I left the place behind, knowing that Butch would probably beat me worse if he found me after splitting.

I wandered for the next several days and nights, until I settled on a neighborhood that offered an abandoned storefront where I could lay my head for a while. A wino took over a corner of the place a night later.

I completely lost track of time, only later realizing that it was on one of those days in this stretch of time that my eighteenth birthday had come and gone.

The stash of money I'd kept after leaving Butch went quickly. I started to panhandle. At the time I didn't think of it that way. But later I realized that's what I was doing. I was what you might call an intellectual panhandler—something I gave a lot of thought to as I stood talking to strangers, feeling the pressure that I was holding them up from somewhere and having to come up with the least obvious way of asking for money. People didn't conclude that I was panhandling because it wasn't usual for a teenager like me to be homeless. Maybe I'd meet them at a bus stop, as if I was waiting for a bus, and I'd strike up a conversation and we'd talk and then I'd casually reach into my pocket to get my change, pretending only then to discover I didn't have any money for the bus. "Hey, you got a nickel?" I'd ask. Usually they'd hand me more—a dime, a quarter.

For a shy kid, I got to be real good at getting strangers to ante up their change.

Every now and then a burst of desperation would propel me to apply for a job. But without an address or a phone and with my lack of confidence, prospective employers didn't see much promise in me. The wino at the storefront tried to give me tips for job hunting. And when he started to talk about me as the son he never had, I knew I had to find somewhere else to crash, that very night. And finally I made the decision that I had avoided until now—I was going back to the old neighborhood.

Excited and wary, I made my way toward the Glenville

area, camping that night on a park bench in a nearby neighborhood. Unable to rest at all, I was experiencing what sailors who've been at sea for long periods call channel fever—when they arrive just off the coast and have to maintain their position until morning when they pull into port for liberty.

Before sunrise I make it to shore and find myself on 105, turning down Superior, down to Parkwood, and past the elementary school, avoiding the Picketts' street, but turning onto the other residential streets—soon passing by William Howell's, by my friend Michael Williams's place, and then Jessie's house.

The next thing I know, the sun is up and the humans of my home planet are stirring. Cars are pulling out of driveways, kids are out in their yards. Mr. Heywood is at his store, opening up. I walk the streets in a swoon, taking in all the familiar sights I've been missing, watching for someone that recognizes me, until, just before noon, I spot Jessie dancing down the street.

"Fish!" he shouts when he sees me. He saunters up and gives me five. "Where you been at?"

I've never felt so happy to see anyone in all my life. Smiling wide, I tease, "Jessie, man, you don't have to say 'at' at the end of every sentence."

"Where you learn that at?" he jokes. "Boarding school?"

"Something like that." Then I tell him the truth, mostly—the thing with Mizz Pickett, the orphanage, George Junior, the men's shelter.

"So where you staying at?"

From my lack of response, Jessie tells me I'm welcome to crash at his house for tonight. His momma has so many kids anyway, he reminds me, she probably won't even notice one more. "Come to the back door at ten," he instructs me, so he can sneak me down to the basement. "You can stay down there."

"Maybe we can get an apartment together."

"You trippin'. I'm staying with my momma."

When Jessie lets me into the house in the night, he takes off afterward for his girlfriend's place. The following day, Jessie makes the same offer. And the day after that. I can bathe at his house whenever I want and he gives me his clothes to wear, too. I stay almost two weeks before taking a break and spending the night at Michael Williams's house. But when Michael's mother comes upon me sleeping on the extra twin bed in Michael's room, she's none too pleased. "Who's that boy in my house?!" she shrieks, waking Michael. So back I go to Jessie's, where I continue to remain undetected by his mother.

Jessie is the only one I tell about what really happened in the time I've been away. Whenever anyone else familiar sees me, the first thing they say is, "Fish, where you been?"

My stock answer: "Out of town."

As good as it feels to be back with my friends, I know that my staying at Jessie's is only temporary and that I have to come up with a plan for myself. Now and then I mentally visit the goals and visions I'd begun to create for myself but they no longer seem remotely tangible. Homelessness, in spite of other people's roofs that covered me for the time being, has made me feel unworthy even to dream.

The only dream I manage to hold on to is seeing Freda again. And yet, because of my embarrassment about my situation, I go to great lengths to avoid running into her. I can't just show up. I have to come up with something to offer her, something to make her see and believe all the promise in me that I can't see or believe.

The more I think about what that offering could be, the more ashamed I become, to the point that I try to distract myself from thinking about her by throwing myself completely into whatever activities I can find. At the outdoor basketball court at Parkwood Elementary, I keep busy with as many games as possible.

I'm in the middle of a game of 21 with a group of guys one day when Jessie stops by. "Fish, where you at?" he calls as I shoot the ball.

"Seventeen," I answer, and head to the other side of the court.

"I'll be at the corner store," Jessie says, nodding in the direction of Parkwood and Superior, where he usually buys his midday drink.

It isn't much more than fifteen minutes later that I finish the game and stroll up Parkwood toward the corner store. Not in a hurry, my mood is light and momentarily carefree as I enjoy the sunny, blue-skied noon and the breezes that have begun to blow here in the waning days of summer. My head is down, my eyes directed just ahead of my feet. As always. Look up, says a voice inside me, look up and turn your head. I look up and turn my head in the direction of the next block, where I need to cross the street to the store.

There, standing in front of the store, I see Jessie, still a young bullish James Brown look-alike, short and strong, facing off with an old wiry man. It appears like a still shot in my mind. All sound and motion cease. Then comes another frozen frame: the sight of the man pointing a gun at Jessie. All of a sudden, I cannot see, except for the blinding white color of the bullet's scream as it explodes from the gun into my friend. In the next several frames, everything speeds up: Jessie's body tumbling down to the side-walk, me running, a crowd gathering, people shouting, street noises wailing, the old man yelling. I'm standing next to Jessie as a pool of blood spills from him. The old man is ranting about how sick and tired he is of these hoodlums terrorizing the streets, how Jessie tried to stick him up and he only shot him in self-defense. Then the police arrive and break up the crowd, telling us to clear out, pulling aside only those who said they witnessed what happened.

Shocked, in a daze, I walk away, following others in the crowd, most of whom probably don't know Jessie. Turning back, I catch my last glimpse of Jessie and then I run, without knowing where I'm going. I run until I can't go anymore, sorrow and loss pulsing through me so violently I think my heart may break.

It is nighttime, in an alley where I'll sleep for the night, when I finally break down and cry. The weeks that follow are a time of constant tears, a return to homelessness, one of the two worst periods of my life—the other being my exodus from the Pickett house. Now it is time, I know, to make a true exodus.

It was Indian summer in Cleveland in the fall of 1977. Being homeless, I appreciated the reprieve in the weather as never before. The nights were getting cooler, but not like the wintry nights of some autumns I could remember. Jessie's death was ever-present in my thoughts, not always in a conscious way, but always there. It showed up in my dreams, in a recurring nightmare of that day on the street that I wasn't able to prevent. I tried to make sense of what had happened. I knew Jessie. He wasn't a bad kid. Not that it was past him to stick up an old man for money to buy wine. But Jessie wasn't walking around premeditating how he could kill somebody. He was just reckless and hot-blooded and not attuned enough to the world to know when to be afraid. He believed he was invincible, and he wasn't. If anything good came to me from Jessie's dying, it was that lesson—a lesson that would help me in the future to better understand the circumstances of my birth and a key to forgiveness of the father who hadn't lived to be there for me.

Jessie's death was also a sobering reminder of Bill Ward's admonitions to me about taking charge of my life. I had been at sea since my departure from shipwreck island, mostly allowing the storms and tides to take me with them, allowing myself to dock, now and then, enough to gather steam to cast off again. Now I needed to chart a course. And maybe if I figured out where the somewhere I belonged was, I'd figure out who the somebody was that I was meant to be.

Before doing that, there were still a few people who were important for me to see. One of them materialized right on the street in front of Cleveland Trust, at the corner of 105 and St. Clair, where the Rexall's with the escalators and the big furniture store had captivated my attention so long ago. I had only just come from downtown after rescuing some of my clothes from a hanger at the YMCA, including the jacket that I needed now that the weather was cooler.

"Twonny!" came a familiar voice from a half a block away. Looking up, I saw Flo. Four years older than the last time I'd seen her, she had changed. Still average build and average-looking, she was looser in her style, much more grown-up. "What are you doing?" Flo asked me with a hug.

Playing down the clothes I was holding, I made up

something about having a place, everything being cool. If Flo was still in touch with the Picketts, I didn't want her giving them anything but the most glowing report.

Not only was she in touch with them, Flo told me in the next breath, she was actually living with Mercy in Bedford Heights, a nice Cleveland suburb, helping baby-sit Mercy's two kids. Mercy. Many nights without anywhere to stay, I'd thought of Mercy's past kindness and wondered how to reach her. "You should call her," Flo said, writing down the phone number for me. "I know Mercy would love to hear from you."

It took me a day to call, but when I did, Mercy insisted I come over immediately. I took the bus that traveled through Warrensville and I happened to be staring out the window when we passed by a recruiting office, one of those multiservice recruiting places where you could sign up for all the branches of the military. At the top of the window was a large sign for the army, the sight of which instantly turned me off and made me look away.

My reunion with Mercy was warm. She, Flo, and I sat down in the living room of Mercy's comfortable town home and made small talk. References to Mizz Pickett were kept at a minimum, although I learned, with relief, that she and Mr. Pickett were long gone from Cleveland, having settled back in the South for some time. Before dinner, the doorbell rang and Mercy told me to answer it. There on the porch was Dwight. No longer an angry teenager, he had transformed into a very handsome young man whose good-looking features were made more masculine by his rugged worldliness. As we talked, I felt that Dwight's unique intelligence was no longer as pronounced; instead, he seemed more cunning in the way that the world can make a person who is forced into it too soon.

Sitting there that night I felt we had a real bond, Flo, Dwight, and I, because of what we'd survived together. I was genuinely glad getting to see them, knowing that we were still connected. But in a silent corner of my mind, I thought how sad it was if this was as far as our lives had gotten us—back where we started. After everything, here we were, still in a foster home.

Echoes of past promises taunted me. "I'm gonna grow up and be a painter like Michelangelo," I'd told Mrs. Profit. "You're gonna read about me someday—for something good," I'd said to Mrs. Brown in the orphanage. "I want to make something of myself," I'd blurted out to my history professor at George Junior. "What I want to be is a photographer."

From what I gathered, Flo and Dwight, each in their own way, had accepted that this was as good as it got. I couldn't. They were so much more adult in their resignation. Maybe I was naive and undeveloped, even delayed, but if growing up meant the death of dreams, I chose immaturity.

Bits and pieces of our stories were stitched together. I tried to put a positive spin on my situation, but Mercy was able to gather that I needed somewhere to stay. That same evening she offered her sofa, where I would camp for most of the next two months. Being so good-hearted and generous, she insisted on giving me spending money, too, in return for my helping out around the house and baby-sitting her kids when Flo wasn't there.

Maybe if this had happened before I had been homeless, I would have been able to feel more comfortable staying with Mercy. But as it was, I knew that I was only delaying the inevitable—either taking the necessary measures to map my own future or, failing that, letting another unforeseen event toss me out into the cold.

These were the lines along which I was brooding one morning while lying on the bed in Mercy's extra room and listening to the radio when the disc jockey announced the lineup of a concert that night at Cleveland's Sports Arena, the same arena where the Cavaliers played basketball. As he reeled off the names, it sounded great—the Bar-Kays, Tower of Power, Lenny Williams. Then the DJ uttered words of magic—the Dramatics! Oh, but there was more. That very day, the Dramatics were going to be at Peaches, a huge local record store, signing autographs and promoting their latest LP, *Shake It Well*.

I'd bought the album a week earlier at Peaches but hadn't been able to play it yet because Mercy didn't have a stereo. In fact, she was getting ready to buy one just so that I could listen to all the music she knew meant so much to me. Setting aside all of my life-and-death decisions, I grabbed the album and my coat and dashed off to see my treasured Dramatics in person for the first time.

Peaches was on Libby Road, only a short walk from Mercy's, and I was early enough to be one of the first to see the Dramatics when they arrived at the store. Unbelievable! Ron Banks, L. J. Reynolds, Lenny Mayes, Larry Demps, and Willie Ford—all five of them standing right there in front of me. I felt like what I'm sure the Beatles' biggest fans must have felt upon seeing them with their own eyes for the first time.

For over an hour, I stood back, *Shake It Well* tucked under my arm, too shy and petrified to push myself into the crush of fans greeting the Dramatics and having their records signed.

When the function was almost over, I became aware that L. J. Reynolds was looking in my direction and signaling me to come on over. No doubt he'd noticed me standing there all that time. Charismatic and regal, with his rich warm voice that could rasp soulful and low or soar into falsetto, L. J. Reynolds was signaling to me! Nervously, I ventured forward close enough to hear him say with enthusiastic zeal, "Hey, brother, how you doing?! Come here, let me sign your album."

Cool, pretending nonchalance, I handed it to him.

As he signed the album, he asked, "Have you listened to it yet?"

"No," I replied, "I haven't heard it yet."

Then he said, "You're going to love it, man. You're going to really like it. It's pretty good." He paused before asking, "You coming to the show?"

Coming to the show? I thought, there's no one to stop me. I told him, "Damn real . . . I'm going!"

"I'll see you tonight." He smiled.

I smiled back and said, "Okay, man," and floated out of the store, out into the day, wondering how on earth I was going to get to the Sports Arena.

The question was answered in the person of Michael Williams, whose house was my next stop and who happened to have a used Ford LTD. No small coincidence that I went to him first. Then he threw in a wrench by saying, "We should take girls."

"Cool," I said, going along with it, as if I'd gotten over my terrible shyness about girls, not letting on that I was panic-stricken inside.

Michael said, "And after the concert we can break 'em down." To my look of confusion, he said, "You know, come on, man, get some poontang."

All I could hear was Smokey Robinson's "Virgin Man" screaming in my head and good-looking, outgoing Michael Williams was like Teddy Pendergrass, the sex expert, singing, "Close the door and let me give you what you been waiting for."

Getting a date for the concert that night?

Suddenly, a bell started to ring. Freda. An invitation to go see the Dramatics with me. It was perfect—the very offering I'd been trying to find to give me an excuse to go see her. Not to mention that she lived around the corner from Michael Teddy Pendergrass Williams. Before my nerves could talk me out of it, I marched over to Freda's house to ask her out. Mona answered the door with the same surprise as everyone else I'd abruptly popped in on, with the same question, "Fish, where you been?!"

"I was out of town," I said. "Is Freda home?"

"No, but she'll be here soon. C'mon up." Mona, whose pretty looks had improved but whose tough, confident air had softened into womanhood in the almost two years since I'd been gone, waved for me to follow her to their upstairs porch, where I kept a constant lookout for her sister.

"Did you hear about Jessie?" Mona asked, wanting to talk about him.

"Yeah," I said quietly, and changed the subject.

Soon I spotted Freda walking down the street toward the house. She was even prettier than I had frozen her to be in my memory. In fact, the closer she came to the house, the more I could see that Freda had grown into a beautiful young lady and she still looked brand-new. Without knowing that I was looking at her, she wore a smile, and her familiar overbite thrilled me beyond words.

The moment she looked up and saw me on the porch, I began to feel so intimidated. Would she really remember me? And if she did, would she still like me? Would she allow me to enter her life again after having disappeared? The answer to all those questions, somehow, lay in the promise of those Dramatics tickets.

Her smile and her tone of voice as she came up to the porch boded well. Looking pleasantly surprised, Freda said, "Antwone"—ah, man, she said my name, the clue to a girl liking me—"what are you doing here? Where have you been?"

"I was out of town."

With Mona looking on, Freda considered this response as incomplete and began to ask questions others hadn't, like where was I staying, what was I doing, and how did I feel about that.

"Freda," I interrupted her last question, my overrehearsed question flying out of me at her, "do you want to go the Dramatics concert with me?"

She and Mona exchanged expressions of shock. Freda coyly said, "Maybe."

And with a knowing smile, I said, "What do you mean maybe?"

Just then Mona offered, "I wanna go!"

Freda shot her a mean look and asked me, "Isn't it tonight?" I nodded that it was, to which she answered, "Yeah, I want to go."

And that evening in Michael's used Ford LTD we were off to the show, he and his date in front, me and Freda in the back. Having used all my strength to ask her out, I had nothing left to fight the overwhelming power of my bashfulness on the date itself. I hardly said two words to Freda on the long drive to the Arena.

We were almost there when she let out a sigh. "You haven't changed at all, have you, Fish?"

I looked at her, disappointment overflowing that I hadn't overcome my shyness at all. "No, I . . . I guess not."

Freda touched my hand as it rested on the seat and said, "It's okay."

The rest of the night was a blur of nostalgia, with the Dramatics performing every song that had ever glued my heart to hers. She enjoyed it as much as I did, but much too soon we were back out in front of her house, standing face-to-face in the cold night air. Without a kiss, we said our good-byes, and I watched her walk onto her porch and into her house.

I didn't know then that the good-bye we had after the Dramatics concert was our last good-bye and that I would never see her again.

Would it have been different if I hadn't failed so badly as a date that night? I'll never know. What I do know is that I had said all the good-byes that I hadn't been able to say when I'd left the first time. So, choosing perhaps the worst season in which to go, with Christmas on the way and snowstorms predicted, I left Mercy's house, plunging myself back into homelessness.

Three days before Christmas I spent the night in an alley downtown, not alone; in fact, there were a lot people huddling in blankets, vying for space. That was a piercing cold I'd never felt before, with the icy winds blowing off the lake and funneled by the tall buildings like a supersonic death-destroying weapon.

In that mind-and-body-numbing state, Bill Ward's old plan for me rose from ashes in my thoughts. The idea of the military made some sense, at last. It couldn't be any worse than this.

The next day, I went to see a friend out in Warrensville, hoping for an invitation to stay under a shelter that night. As if by coincidence, though not at all by chance, the bus let me off on the same street as the recruiting office. I came closer to it, cringing all the way. But then my eyes moved away from the army, air force, and marines posters over to another poster in the window. I had passed by this recruiting office often in the past two months, but this was the first time I'd noticed this poster.

That was the poster that lured me inside. It directed me to the attention of two recruiting officers who may have assumed correctly from my disheveled, frozen appearance that I was homeless; who also assumed by my young looks that I wasn't telling the truth about being eighteen; who only believed me after I mysteriously produced both my birth certificate and my high school diploma.

Before I crossed the threshold, I stood and gazed at the poster at length. It was an old-fashioned movie-style poster with the word *Heritage* printed in large ornate letters at the top. In the background was a wooden sailing ship, and in the foreground was a sailor, dressed in a crackerjack uniform, holding the hand of a boy as the two gazed off into the distance together—toward adventure, fun, and glamour. And at the bottom, in bold, inviting type, was the phrase that really got me: *Join the Navy, See the World.*

Antwone, he's a dreamer, he'll dream his whole
life long. Antwone broods over love that hurts and
love that has gone wrong.

Antwone loves tomorrow and hurts for yesterday;
Antwone hopes to find a love that will love and
always stay. All he wants is true love, seems not so
much to ask. To find this love has proved to be
Antwone's greatest test.

Antwone loves so gentle, a way words cannot
describe. He's so sentimental, he sometimes . . .
well, he cries. To many he's a mystery, with eyes that
cannot lie, a mystery—oh, I don't know, I think
that he's just shy.

Antwone has traveled the world, and many know
his name. If you're surprised, he'll just sigh and
quickly call it fame . . . so vain. Antwone is
romance; he's also kinda cute, Antwone loves a
candlelit night and dinner just for two. Antwone
is a fashion of charming debonair, a warm
enchanted fellow that no woman would ever share.

Antwone loves the rain, and times love brings in
spring. Antwone loves to dream of joy, you know,
that kind of thing. Antwone is a dreamer. I'm sure
he will always be. I think I can say I know him
well, for Antwone . . . well, he's me.

MAN OF THE WORLD

seven

Get your ass over there and do like I tell you!" barks the chief boatswain mate, his voice roaring even above the massive churning diesel engines of the U.S.S. *Schenectady* LST 1185, and above the loud chopping sound of the incoming Hueys—the huge two-prop marine helicopters lining up to land on the flight deck.

In my past months home-ported in San Diego as a seaman apprentice assigned to Deck Department, where we were teasingly referred to as deck apes, I haven't flinched from a single assignment. But now the chief boatswain mate is asking me to do something I'm sure is far beyond my capabilities—to be the sole person responsible for guiding the pilots to a safe landing. Already, the first Huey is approaching as I'm arguing with the chief, a guy who looks too much like Elmer Fudd for my comfort.

From the first night that I arrived at boot camp in Great Lakes, Illinois, two nights before Christmas, I've had people barking orders at me, determined to drive out all traces of civilian blood from my body. It began the instant we stepped off the bus into what felt like arctic cold and met up with the rest of our company. Out of the dark came a chorus of orders to march. Eighty of us—black, white, Asian, Hispanic, from many walks of life, all

285

young men—proceeded to march, each in a different timing, everybody bumping into one another, stepping on each other's heels. Total havoc.

"Anybody here ever been a Boy Scout?" yelled a petty officer second class.

I raised my hand, and he asked me to demonstrate how to march. I had not gone two steps when he yelled, "Take your ass to the back of the line!"

Like Keystone Kops, we marched to our company living quarters where double bunks lined the hall—our home for the next thirteen weeks. At 3:30 A.M., the lights were thrown on and a large garbage can was tossed down the middle of the aisle. Bolting upright out of sleep, I heard the shout, "Drop your cocks and grab your socks!"—a phrase I came to know well that means it's time to get up.

Our days were exhausting, filled with instruction as basic as how to properly press and fold every article of clothing we wore, down to our briefs, to detailed academic courses, to pounding physical training. We were even taught prayers that had us saying good night to the navy's various patron saints, like Chester Nimitz, thus teaching us some additional naval history and loyalty.

Early on, we learned about the different ranks to which we might one day ascend—from Recruit, at the bottom, all the way to the top, President of the United States, the Commander in Chief.

Every day we were given inspections and tests. If you failed any of the inspections, if one T-shirt wasn't folded to absolute perfection, you were given MTU—extra military training that required running through an obstacle course set up in a massive drill hall while you held a rifle aloft over your head. Any kind of mild infraction of rules, including falling asleep at any time during the day, would result in MTU. Like everyone, I feared and dreaded MTU. Every inspection and every opportunity to do something wrong, I was sure that my number was up and I'd be sent to the drill hall. But not once did that happen.

Failing any of our weekly tests resulted in something far worse than MTU—being sent back a week. Here the navy really had me scared. Being sent back wasn't just having to take the course over again, it was being moved to a different company of recruits who had begun later than us, and to be in boot camp that much longer. Eventually, if you were sent back enough times, you were deemed unqualified to serve and given a discharge, even though you had never made it into the navy.

Out of eighty recruits in my company, twenty or so were eventually sent back or discharged. Whenever any of them failed a test, their names were called afterward and they were told to wait outside. We never saw them again. Through every week, through every test, I listened with a clench in my gut, sure that the next name asked to wait outside was going to be mine. It never was. I passed every test.

Up until now, with the exception of Mrs. Profit's class and my popular period in junior high, I'd been below average. And by no means had I ever been an athlete. But boot camp gave me a new self-image. I found out that I was average, academically and athletically. In some instances, I was even above average, especially when it came to cleaning and organization, as well as adhering to rules and regulations. If my upbringing in institutions had prepared me for anything, it was definitely those aspects of the navy.

I may have also had an easier time in the way that boot camp sought to eliminate our individuality. After all, my individuality had never been allowed to flourish before.

And I didn't suffer like most of the other homesick recruits who existed for daily letters from home and girlfriends.

Not that there weren't days when I thought I couldn't do it anymore. But when faced with quitting, as a few in my company did, I knew that wasn't an option for me. If I quit, I'd be right back on the street, homeless again. For now, grueling as it was, boot camp was my home, the best place for me.

And, indeed, toward the end of the thirteen weeks my company became my family. We were Company 902, a number that held special meaning for me since it was the reverse of Mrs. Profit's classroom number, 209.

The navy had powerful yet subtle techniques for creating company unity—the kind of loyalty to one another required if we were to go into battle together. There were tests given, for example, that every recruit in the company had to pass. If one person failed, the whole company failed. With us, that almost never happened. By the end of boot camp, Company 902 had won every competition, every flag and award. Then on one of the last drills in PT, a small guy in our company collapsed with a leg cramp and couldn't complete the run. Because of our outstanding achievement so far as a company, our company commander was able to grant permission for him to graduate with us and not be sent back. When the commander announced the decision, the silent cheer that went up made you think it was everybody's ass that was in jeopardy.

Company 902 carried state flags and marched during other recruits' graduations. These took place in the main drill hall, a cavernous place built like an airplane hangar, big enough to accommodate two thousand marching recruits. During one event, our company member assigned to give us the signals for when to raise the flags, when to lower them, and when to turn, suddenly went blank. I stood there frozen, holding the flag of New Mexico upright, knowing that he was supposed to give the command to lower it. When he didn't, not wanting to bring attention to his mistake, I decided not to lower my flag, even though it was what we were supposed to do and not doing it could get me into trouble—like being sent back a week or two. Not one of us lowered our flag. Our company commander later remarked how fascinating it was that sixty of us made the same mental decision. We were thinking as one.

Our own graduation ceremony marked one of the high points of my life. Until that day, I had seen myself as a lone figure in the world, not connected to anyone, small and battered by the elements of circumstance. But that day, when I put on my dress blue uniform—in the admiral style, navy blue wool with silver buttons—I was amazed to see how solid and strong I had grown. At six feet, I was tall, broad-shouldered, and long-legged, my physique well shown in the tapered navy jacket and straight dress slacks. And as I marched into the drill hall amid two thousand other recruits, I was connected to them all. Moving in seamless formation with my fellow company graduates, I flashed on our fumbling attempts to march a mere thirteen weeks earlier. What a transformation we had made. We sounded like the approaching footsteps of one big giant—*march, march, march, march*— every foot in step with every other, the line cutting and turning from side to side in perfect sync. Not only that, but as I stood at attention and heard our company commander praise us, saying with true passion, "There will never be another Company 902," I realized that no one could tell my situation was any different from that of anyone else here. That was what the navy did for me from the start. It made me belong, it made me feel that I was the same as everyone else.

One of the company bands, made up of recruits who were selected for their musical aptitude, began to play a solemn ceremonial hymn and soon I heard my name announced as a graduate and official recruit of the United States Navy. A purity of feeling washed over me—pride. I was Antwone Quenton Fisher and I was proud to be me. I was proud of myself and the place where I now belonged. If only somebody could see me, I thought; somebody should see me. Who that somebody was, I didn't know, but I was wishing—if only anybody who knew me before could see me now.

A couple of days later, dressed in my uniform and pea coat with the entire contents of my life to come packed in the seabag on my back, I arrived at Chicago's O'Hare Airport and headed toward the boarding area to take the airplane to San Diego. I turned to look back and watch the other graduates dispersing in different directions. Our lives were before us. My life was beginning. Not another chapter in my sad history, but another book, another life entirely, just starting out, with every possibility of every good fortune ahead of me. For the first time, I was the kid I had never been. And for the first time ever, I knew what it was to feel new.

Even though boot camp had given me a solid foundation for the navy, there was so much more I had to learn, something I discovered during my first day of work in the Deck Department. At the time of my arrival, it so happened that the U.S.S. *Schenectday* LST 1185, a troop landing ship that could physically pull up onto the beach and open its bow to deploy marine tanks, was in the midst of an overhaul, having recently returned from a cruise of the western Pacific. YB, the leading seaman in charge of us my first day, informed us that our job was to paint the starboard bow.

With zest and vigor, I took my position on the small floating barge and began to paint with tremendous dispatch. As I recalled well from my younger days, working hard and fast was the way you improved your station. In fact, I was moving so quickly and thoroughly, the other guys began to grumble. "Hey! Hey! What's your name?!" asked one guy.

"Fish," I answered, continuing to paint.

"Fish, what you tryin' to do? Slow that shit down!"

"We gotta make the job last till knock-off!" said another.

Before long, YB called down to me, "Fisher, slow down! The way you're going, we'll be done by lunch."

"Don't we want to finish?"

"No, because if we do, they'll only find something else for us to do. Take your time."

Slow down? Okay. What an eye-opener. Clearly, it was important to do a good job. But the main idea today was to make the job last. This was one of many times I would see that some lessons I learned about survival in my childhood had become obsolete.

After our overhaul, the *Schenectady* was elected to assist in qualifying marine helicopter pilots to land on a ship in motion. Initially my job on that relatively small deck landing area was as one of four chockers who stood in pairs to either side of the man whose job it was to give hand signals to the incoming helicopters. Once the helicopter landed, we ran together, two to a side, and secured the helicopter's wheels to the ship with chains and blocks, also known as chocks.

After three days of chocking, Chief Elmer Fudd decides I'm ready for the most important job on the flight deck—guiding the helicopter into landing. After some training, here I am in the brisk ocean spray coming off the high seas, with a battalion of Hueys heading toward the ship, getting the cursing of a lifetime because the chief thinks I'm ready and I know I'm not.

In theory, I have it down—waving the pilot toward me with one hand, waving two hands to the side for when to stay in position, then waving toward me again. But in practice, I can just see myself freezing up and forgetting the proper signals and sending the Huey and its pilot crashing into one of the ship's tall stacks, killing and maiming everyone.

"You don't understand!" I argue with the chief. You don't know me, I want to say, how can you put everybody's lives in my hands?

As I back away from him, Elmer Fudd takes off his helmet—his brain bucket, he calls it—and throws it at me, hitting me in the chest. "You hear me? Get your ass over there and do what I tell you to do!"

With the chockers and everyone else in position, I nervously take my position and look up in the sky. The Huey doesn't look so big over the ocean, but the closer it gets, the bigger it grows, swooping toward me like some gigantic predatory bird. From other positions, these helicopters are loud, but there is nothing like the sonic-barrier-breaking sound of being directly in its frontal path.

As the Huey gets closer still, bigger still, louder still, I look up to see the shadowed face of the pilot trained on me. With his headgear, his features are hidden in a dark, ominous mask, but I can tell that he is watching me and waiting for my direction. The chief and everyone on deck are watching and waiting. And then I'm doing it, I'm bringing him in, convinced still that it's too much responsibility for me, that I'll send him off course, because my luck is not for it to work out, my luck is for this to be the worst day of my life. I freeze again as the pilot hovers just above the flight deck and the chief explodes from the side, "Goddammit, Fisher!!"

A jolt of electricity pulses through me as my arms move, correctly, decisively, and I give the signal, the right signal, and the helicopter lands, bounces a little, and stays. Just like that. I signal for the chockers to secure the Huey to the deck and they do.

The chief walks over to me and nods his head. "See? You can do it. Now let's see you do it again."

For the next hour, I guide a dozen Hueys to safe landings and takeoffs. The more I do it, the more confident I am. And when the chief tells me it's somebody else's turn, I don't want to stop. I can't wait until the next time I can do this job and experience this unforgettable feeling of power, purpose, and importance that springs from the realization that everybody was depending on me and I didn't let them down.

In the book of my new life, I am not only able to be responsible for myself, but I can be responsible for others—for their lives. In this one day on the deck of the U.S.S. *Schenectady,* from the goading-on of a chief boatswain mate who reminds me of Elmer Fudd, I meet and greet my powerful self, thus leaving childhood behind and being born a man.

I served for eleven years in the United States Navy. When the U.S.S. *Schenectady* returned from a lengthy overseas cruise, I opted not to stay on the ship in its home port but to transfer to another ship that was leaving out of San Diego for a long cruise. My feeling was that I had joined the navy to see the world and I was intent on seeing as much of it as possible. This became a policy of mine throughout my service. Fortunately, it was easy to find sailors on other ships willing to swap duties with me, since many of them, because of family or girlfriends, didn't want to go overseas.

My second ship was named, ironically, the U.S.S. *Cleveland* LPD 7. It was also a large troop-carrying ship. From it, in 1980, I was sent to Pearl Harbor, where I served on two ships, the U.S.S. *Edwards* DD 950, a destroyer, and the U.S.S. *Robert E. Peary* FF 1073, a fighting ship called a fast frigate. For part of my time in Hawaii, I also served at the Naval Communications Station Wahiawa on the island of Oahu. Then I was assigned to the U.S.S. *St. Louis* LKA 116, which was based in Sasebo, Japan. Coming almost full circle, I completed my last two years in the navy in Long Beach, California, on the U.S.S. *Mt. Vernon* and on shore at the Naval Memorial Hospital.

In the course of my eleven years, on every ship, at every station and every stage of my development, the navy provided me with essential lessons like that one I learned by landing Hueys. I didn't always recognize the lessons at the time, nor did I always appreciate the teachers and their teaching methods. But I was still paying attention.

Senior Chief Lott was a case in point. A chief boatswain mate, Lott was a black powerhouse of a man with a mouth so foul I was sure the expression about cursing like a sailor had to have been invented for him. He transferred onto the U.S.S. *Schenectady* before we left on our first cruise, and his loud swearing could be heard morning, noon, and night throughout Pier 9 at the 32nd Street Naval Station in San Diego.

Chief Lott was such a taskmaster that a senior officer, an ensign, took pity on us one day and told us to knock off ship's work for a couple of hours early, at 2 P.M. instead of 4 P.M.

"Fuck that shit!" blasted Chief Lott. "He may have the bars, but I got the stars. Your fucking asses is gonna be working till dark thirty!" Translated, that was thirty minutes after sundown—meaning two extra hours. When we complained, he let go another barrage of expletives.

Every day at sunrise, Chief Lott was ready to go with more four-letter words, which would echo throughout the quiet harbor. Officers from neighboring ships complained to their captains, who complained to our captain—who, one evening, summoned Lott to address the problem.

Imagine our surprise the following day when Chief Lott appeared on deck for muster instruction and inspection, smiling demurely. "Good morning, ladies," he began, gushing. "Good morning, sweethearts." As my jaw fell open, Chief Lott addressed me. "Why, good morning, Miss Fisher, would it be too much trouble to ask you to close your mouth?" Batting his eyelashes, the chief turned to Seaman Wilson, put his hand on his hip and purred, "Why, good morning, Miss Wilson, you sure look lovely this morning."

He kept it up for the remainder of the day, all the way until knockoff. It got to be more annoying than his cursing. Whatever intervened, we didn't know, but the next day Chief Lott was back to normal, with his regular early A.M. greeting: "Fuck you all! Get your fucking asses moving!"

It wasn't Chief Lott's NC-17-rated mouth that bothered me so much, it was the fact that he began to ride me with it constantly. His first big gripe was about how I walked with my head down. He took it as a personal insult, unloading on me one day after knock-off as I started to go to my berthing quarters. "Fisher!" he barked.

I turned, as did the other guys still in the vicinity. In my early months of the navy, I was known as easygoing, quiet, and hardworking, and it was rare for me to be called out. But clearly, from the hard look on his face, Chief Lott had it in for me about something. "Fisher," he said, "I'm sick and tired of your fucking shit. I'm tired of looking at you walking around with your knucklehead down." Aware that the other guys were watching, I bristled. But before I could move or say a word, he yelled, "So lift your fucking head up!"

I raised my chin.

"I don't want to fucking see you with your fucking head down again."

"But—"

"Don't let me fucking catch you," Chief Lott warned. "Don't let me walk up behind you, don't let me drive up next to you, don't let me pass beside you, don't let me walk up adjacent to you and see you with your head down! Because if I do"—he paused for a long moment—"well, just let me catch you and you'll find out real fucking quick!"

I started to object, "*Damn,* Chief—"

Chief Lott froze as if in shock and looked at me with a quizzical tilt of his head, like a dog that's picked up a potent scent. "Fisher," he said, his eyes like slits, "are you cursing at me?" Answering his own question, he announced, "Yes you have, you just cursed me out! You'll pay for that, you'll work for that fucking infraction."

Taking his threat seriously, for the next several days I looked for him around every corner, making sure that I had my head ready to lift up the second I sensed him coming. I became so paranoid that I decided I just wouldn't have my head down at all. Before I knew it, I was walking around the ship staring up at the sky most of the time, like a person craning his neck to see something on a high shelf.

But that wasn't enough for Chief Lott. His next bone to pick was my soft-spoken, careful way of talking. He seemed to understand nothing about my years of nervous shyness and the old stutter I tried to cover up with a low voice. Again, he acted like it was a personal affront to his senses. "Fisher," he charged, as I stood there at attention looking at the space above his head.

"Yes, Chief?"

"I can't hear you, Fisher!"

"Yes, Chief!" I tried to shout.

"Just fucking speak to me."

"Chief Lott . . ." I began, fighting off the anger.

"What the fuck is wrong with you, Fisher? Speak up when you talk to people. Direct your voice and direct your attention *to me*. Talk *to me*. Lift your head, direct your voice, look at me. Speak to me!"

I had no choice but to do as he said. It felt like I was some bizarre contortionist, my neck craned up like that, my voice a trumpet blaring out from inside of my throat, right at him, my eyes pulled tight into slits. And I tried to be cheerful doing it. "How ya doin'? What's goin' on, Chief Lott?"

"Fisher, goddammit! What the fuck is wrong with you?"

Not only did he want me to do all those things, but he also wanted me to do it as if it came naturally. Soon, I not only had to lift up my head and direct my voice and attention at the person to whom I was talking, but also do it in an authoritative way. To do that, Chief Lott insisted, "Lean in when you talk. Speak *to me*."

"I'm talking to you, Chief."

"Goddammit, Fisher! Talk to me. Lean in. You're a man, ain't you?"

The whole time I struggled with his instructions, I hated his guts for grilling me in front of others; apparently, I thought, he hated mine, too. To the contrary, I later discovered, Chief Lott cared a great deal for me and wanted to help me rid myself of the burden of my own show of inferiority. Later I would see that he was exactly the right teacher for me, in my life at exactly the right time—giving me the very communication skills I needed to help me advance in my naval career and in life.

Without the added confidence that came with my new posture and manner of speaking, I might never have accomplished my main goal: to get the hell out of Deck Department. One guy had already done it and I was determined not to be the last one left standing, or swabbing, as it were. The way out of Deck was to choose what job you wanted from among the many different jobs around the ship, take the study courses required for it, and then apply. The job I chose, not for any love of it, but simply because it was to me the cleanest, was Ship Service, the department that handled the barbershop, the laundry, the vending machines, managing the bulk storerooms, and operating the ship store.

In order to make sure I got this job, I went all out. After knock-off, I'd go to the laundry and help the ship servicemen wash and press the clothes, even to the point that they'd give me the key and let me do the work on my own. In my other spare hours, I trailed the ship's barber, watching how he cut hair, sterilized his tools, and maintained the appointment log. Then I began to cut hair. In the beginning, my expertise was cutting the black guys' hair. But before long I branched out to everybody—

straight hair, curly, fine, thick. Everybody's hair is different, I discovered; the trick is that you have to see the way it grows before you start attacking it.

I became so good, in such demand, that I never had to do the six months of scullery—kitchen detail—that everyone else was required to do. While on the U.S.S. *Cleveland,* I impressed the captain of the ship so much with the way I cut his hair that, upon his recommendation, Rear Admiral Ramsey came to our ship specifically for me to give him a haircut—an event important enough to be acknowledged in my record.

This was the file that, interestingly enough, the navy kept about my progress through its system, not unlike the official document kept of my years as a ward of the state. But unlike the entries from my childhood that described me as retreating, inexpressive, and functioning at a below-average level of intelligence, my superiors in the navy went so far as to call me a "young intellectual" who

```
 . . . performs duties in an orderly
and professional manner. Has a force-
ful attitude that gets things done.
Very impressive in uniform and
civilian attire.
```

I may have appeared impressive to some in my off-duty attire, which was always dark hues of brown and black, but to SH1 Slikowski, it was a sign that I needed to lighten up outwardly and inwardly. Of Polish descent, short, with a wide, muscular torso, small waist, and washboard stomach, Slikowski walked around the ship like Hulk Hogan. He somehow managed to doctor up his uniform so that it fit him like a glove, showing off his impressive physique.

On my way on liberty one afternoon, SH1 Slikowski stopped me on the quarterdeck and asked me why I always wore dark clothes. Until he mentioned it, I was unaware that I dressed so somberly. Slikowski suggested I buy some lighter hues. "Put some color into your life, Fish! It'll brighten your mood."

Taking his advice, the next payday, I went and bought myself a few new items of lighter colors—mostly white and pale blue. Sure enough, the new attire did seem to lighten my mood and my feelings. Over the years, I remembered his advice and the simple idea that changes on the outside could make you feel different on the inside. It became a habit, now and then, to reinvent myself when needed by tossing out civilian clothes I'd worn long enough and replacing them with new ones. I even found myself passing on the advice to others I saw falling into ruts: "Put some color in your life!"

Later, when I went to work in the civilian world, I found occasion to quote Chief Lott to others, too, especially when they wanted to give me an ass-chewing. I'd tell them, "That's not the way to do it. I've been chewed out by the best. You got to lean in. Talk *to me*. Lean in and speak directly *to me*."

During my first two years, while I was growing at a rapid pace, I was also shedding an old skin that came from a buildup of anger that I had never dealt with. Most of this took place on the U.S.S. *Schenectady,* where my medical records showed my deepening concern for what I was going through:

Patient was seen this day at his own request in regard to mood swings and depressive moods and suicidal ideation. . . . Patient currently shows a lot of strength: no overt sign of depression. Able to laugh appropriately, alert, rational. Patient wished to have psych. consultation to ventilate feelings.

The visit didn't change much for me as depression gave way to rising frustration. It was probably what my social worker Patricia Nees feared when she warned that I was a walking pressure cooker in need of therapy to help me from exploding in an unprotected environment. She understood, as I later did, that anger doesn't dissipate over time; it builds. I was lucky. Rather than exploding, my anger seeped out, often in periods of intense introspection, other times in sudden bursts usually triggered by someone or something that was said to me. Fortunately, that seeping out didn't take place in an unprotected environment. This was the navy, after all, which had a structure for dealing with outbursts of my kind.

The fact that I became a formidable fighter may have seemed incongruous alongside the fact that I was generally well liked, known to be fair, friendly, and good-humored. And I was all those qualities with most everyone—as long as you didn't rub me the wrong way. Call me a racist name, show me disrespect, or give me a condescending attitude, and I had another side that came roaring to the surface.

The primary punishment for infractions such as mine was restriction—a period of time during which you couldn't leave the ship, you were given extra duty, and a financial penalty was taken from your monthly pay. I didn't like restriction, but because I hardly left the ship anyway, the threat of it didn't often deter me when I entered into a fight.

My anger was expressed not only by fighting or running my mouth, but also in subtle ways. If somebody showed up for a haircut and got smart with me or I wasn't in the mood to cut their hair, I'd give them what they called a navy regulation haircut—one that far exceeded the standards.

"Hey, what are you doin'? What'd ya do to my hair?" cried Osanko, a Filipino guy who had the nerve to remind me that I had to cut his hair whether I was in the mood or not. Upset that I took his hair down to the bare metal, he promised, "I'm gonna show the chief what you did to my hair."

That never bothered me, because I knew the chief encouraged me to give the entire crew the same haircut. In fact, he later complimented me on Osanko's haircut.

My regular sparring partner was a fellow ship serviceman named Ivory, a skinny, tall, dark-skinned black guy who, due to drinking in his off-hours, seemed older than his years. He was my same rank, so I didn't appreciate how he tried to order me around. If he told me to handle the laundry and it was his job, I'd tell him that. And if he gave me lip, which he usually did, I'd jump on him.

My archenemy was Istad, a white surfer-dude type with acne that bled all the time, who could often be found popping pus out of his face onto the mirror in the head—and leaving it there.

I was shaving in the head one day and used cotton balls and astringent on my face afterward. Some of the cotton fibers clung to my face.

"Fisher," Istad asked me, "what's that on your face?"

I wiped my face with the palm of my hand and saw what it was. "Cotton," I said.

"What's with you people and cotton?" Istad asked.

And I jacked him up for his comment.

After the navy, Istad later surfaced as an inmate at a California penal institution where my close friend Eric White, a fellow shipmate, went on to work as a prison guard. In the navy, it was Istad's lackadaisical and disrespectful attitude that really infuriated me. One night everything came to a head, literally, when he was supposed to relieve me for watch and didn't, even after I'd gone to wake him to give him a fifteen-minute warning.

"You gotta get up, Istad," I said when he refused to budge from sleep.

"Get the fuck away from my rack," he rasped.

I went back up and when he didn't show for another half hour, I went downstairs and found him leisurely putting his tie on. My reaction was to grab his neck and head under my armpit and ram his head into a fire station—the iron plug where the fire hose is stowed. The revenge was sweet, but my punishment was swift—thirty days restriction and extra duty and a loss of two hundred dollars from my pay for each of the next two months.

That cooled me off for a while, but there was still more seeping to come. Sometimes, I fought for the smallest of reasons, whatever piqued that reservoir of unspent anger. The other sailors didn't mind watching a good fight, but they knew they could get in trouble for being party to an incident and usually they'd pull me and the other guy apart before blood spilled. But once the crowd thinned, I'd dive back in. Miraculously, my superiors didn't find out about these fights. Again, I was very lucky—because whether I acknowledged it or not, there were serious consequences to face. The navy was keeping score, and any kind of label you earned—such as troublemaker—usually stuck for good.

Part of my good fortune in avoiding more extreme punishment was the quality of my work. The navy gave me a few second chances. The last chance was Lieutenant Commander Williams, following an incident while I was aboard the U.S.S. *Edwards,* stationed in Hawaii. Rather than physical fighting, by this point my abuses where more verbal. In this instance, I was being reprimanded for giving the supply officer a contemptuous look.

When incidents took place, usually, a lieutenant JG would investigate the circumstances and put the individual on report; then the individual would be sent on to the XO—executive officer—who would decide whether or not his infraction was serious enough to send him to a captain's mast, like a small court hearing, where the captain would determine the punishment, if any.

Up until now, the XO had seen me for a handful of incidents and had been protecting me. In order to avoid sending me on to captain's mast, he'd been punishing me himself with extra duty and the like. He couldn't do that anymore, and he was at the end of his rope with me. Looking over my record of accomplishments, contrasted with my record of infractions, the XO shook his head in confusion, asking, "What's going on? What's the problem?"

I began to tell him generally about where the anger might be coming from. But realizing I had a lot to say, I finally blurted out, "I think I need to talk to somebody."

He agreed to arrange for that, thus allowing me to avoid captain's mast this one last time.

Lieutenant Commander Williams, a navy psychiatrist, had an office at the Submarine Base Pearl Harbor. In my fairly hostile state of mind, my first impulse was to clam up. But this was my last chance. And the navy wasn't offering me therapy. This was one visit, for one evaluation.

I'm not sure what I was expecting, since my most vivid memory of a psychiatrist was Dr. Fisher, whom I'd expected to be my father but wasn't. Commander Williams's office, cluttered with files, dust, diplomas, navy awards, model ships, and ships in bottles, looked nothing like Dr. Fisher's spare clean space with toys and an impressionist painting that reminded me of the rain. Nor did Williams look anything like Fisher. He was black, for one thing, a large, bespectacled, middle-aged man who seemed worn down by the navy. And unlike Dr. Fisher, who was warm and involved, the commander had a much more serious, distant manner. But the thing that really drew my attention was the way he was dressed—like an absentminded professor, his ill-fitting khaki slacks bunched up in his butt and his white doctor's smock too short in length and in the sleeves.

Holding my file, he led me to his desk, took his seat, and pointed for me to sit down across from him. Then he proceeded to read the reports, ignoring me for several minutes. When he looked up, it was as if he'd only just remembered that I was there. Like an afterthought, he began to rummage for a pen and a writing pad. I sat stoicly, without reacting, even though I had the desire to laugh. But when he started to ask routine questions for the file—date and place of birth—still in his clinical, detached way, and I heard myself beginning to talk about my childhood, I started to feel something else. It was the feeling that I had years before when an uncontrollable urge to cry had come over me. Only now I knew how to hold the floodgates.

It surprised me that I was talking so openly. But it occurred to me that this was going to be like a one-night stand. Why not let it all hang out? And I'd be gone in the morning.

Commander Williams said little, only interjecting now and then with questions. His lack of emotion might have made me think he wasn't interested, but I could also see he was listening to me intently. By listening, he was giving me, as the Indians say, big medicine. Before that meeting with Commander Williams, I had never told anyone my story. I had never been given the chance to connect the dots of my existence, to see the shape and the course of my life, to observe for myself how everything that had happened had its reason, its lesson. To talk was liberation from the prison of silence, from the burden of my own secrets. But by the end of the hour, I had only begun.

Lieutenant Commander Williams knew and understood all of that and decided to give me the opportunity to continue talking. Since the navy didn't authorize him to offer ongoing therapy of this kind, he could only see me officially for one more visit. After that, we met unofficially and sporadically—for walks, coffee, or in the bleachers at a local ball field. Telling my story over a period of almost two years, in this fashion, sometimes with long stretches out at sea in between our talks, I had time to release my anger, slowly and constructively. And when I had insights—sitting by myself in the evening calm, in that quiet time after knock-off when I and many of my fellow sailors would find different secluded spots on the ship in which to be alone to think or write letters or commune in small groups—I could look forward to sharing my thoughts with Commander Williams.

And he was also someone with whom I could share my accomplishments, someone who knew enough about the distance I'd traveled to be impressed. In this time frame, Commander Williams wasn't the only one cheering me on. There was also Senior Chief Akiona, of Hawaiian descent, under whom I served when I was at the naval communications station, where I was assigned to be assistant manager of the enlisted men's club. Chief Akiona called me Tony—don't ask me why—and he decided—don't ask either—that I was going to be the star of the annual PT test the navy required everyone to take.

He kept saying, "Oh, Tony, I know you're going to come in number one."

You don't know me, I wanted to tell him. Number one meant nothing to me. I just wanted to pass. Besides, I hadn't been in training. But the chief kept hounding me about how well he knew I could do. On the day of the exam, he was on the sidelines of the one-and-a-half-mile race, yelling and clapping louder than anyone, "C'mon, Tony, you can do it. You are number one, Tony!"

Out of his sight, I slowed to a walk, but when I'd come around the running path next to him, I'd speed up. But somehow, passing by him and seeing his enthusiasm, I saw that he really believed I could do it. I couldn't let him down. He believed, so I gave it my all. Lo and behold, I came in second, beaten out only by a young woman who was probably a natural-born athlete. The experience made me wonder what it would have been like to have had a father to root for me through life like that—to believe in my ability to win with everything. For that day, Chief Akiona was my father, just as Chief Lott and SH1 Slikowski and others had been in the past.

With all this encouragement, I continued to progress, soon becoming manager of the enlisted men's club. In charge of overall operation, I handled the daily activities report, the cash change funds, overseeing food service and sanitation, managing civilian and military personnel, along with vendor relations, booking bands for entertainment, and ensuring compliance in all these areas with navy regulations. My performance report from this time showed a long, glowing list of credits:

> . . . Petty Officer Fisher has per-
> formed duties in an excellent man-
> ner. . . . He has greatly improved
> the habitability of the club through
> purchase and installation of such
> items as mirrored tiles, ceiling
> fans, and attractive pictures. . . .
> By using cost control formulas, was
> able to increase the daily profits
> in the food service area by 50%. . . .
> Helped plan and organize command
> Christmas party. . . . Instrumental
> in budgeting and booking excellent
> bands and DJ's which have improved
> the morale of the club patrons. . . .
> As warehouse custodian, established
> a good procedure on inventory control
> to ensure no spoilage of food. . . .
> Attended and excelled in Japanese
> language class at Academia School of
> Language Hawaii.

Commander Williams was proud of my progress.
Toward the end of my time talking to him, he invited
me to join him, his family, and close friends at their
home for Thanksgiving. Sitting down to a turkey dinner
and all that goes with it, I was nervous and shy, but so
happy I couldn't stop smiling. Until that visit, I had
spent most every holiday in the navy either by myself
on the ship or working, not taking time off for holidays
or leave.

On the base, Commander Williams was called Mr. Discipline, but that Thanksgiving I saw him in a different
light—more relaxed, less serious. It struck me then that I

knew so little about him and about what motivated him to go beyond the call of duty to listen to me. Maybe he saw me as an opportunity to practice what he went to school to do. Maybe he took pleasure in bending a few navy rules. Whatever it was that moved him, I was grateful.

Besides giving me his ear, there was one thing that Commander Williams said early on that made a huge impact on me.

In telling him the story about who I was, I had accepted that my life began as a foster child. It was along the lines of how black American history is often told, that when we show up in the story we are slaves. To get me to understand that I had a history before my life as a foster child, Commander Williams put it to me in the most basic way possible. "Have you ever considered that you come from somewhere?" he asked, following up his question with a statement. "Because, by the way, you do come from somewhere. You may not feel like you have a mother, but you do. Everybody does. Everybody comes from somewhere."

Try as I might, I couldn't argue. I accepted that I came from somewhere and that someday I should unravel the mystery of where that was.

Before I could do that, I still had other areas of my development in which I was making up for lost time—including my growth as a creative person. As if by seeming accident, early on I made a life-altering discovery when a letter was sent to the captain of the U.S.S. *Schenectady* by Seven Seas, a store that offered a variety of goods and special services for the many sailors stationed in San Diego. The letter said that I owed them money on an item they claimed I had charged to my account. Seven Seas extended credit to sailors with the implicit knowledge that the navy would get involved if bills were paid late or missed.

When the captain brought the letter to my attention, I objected heartily, letting him know that I had paid for the item in the store; and, if he wanted, I could produce the paid receipt. The captain took me at my word and considered the matter closed. But I was outraged. Instead of a physical fight, which wasn't a realistic option, I decided to express my anger in words by writing Seven Seas a letter of complaint. To do so, I got myself a dictionary, paper, and a pencil—none of which I had in my possession up until this point—and began to write: *"It is hard to conceive . . ."*

Conceive was a word I had only recently discovered in a song called "Spirit" by Earth, Wind & Fire, which had the lyric, "It's hard to conceive love's ecstasy." I had been trying it out in conversation lately in the barbershop whenever someone assumed something about me that wasn't true, as in "How could you conceive of saying that to me?" The word was perfect for my letter: *"It is hard to conceive that an establishment as old and respectable and responsible as yours could have made a mistake of this magnitude. . . ."*

I continued in that vein, passionately protesting their heavy-handed methods and enclosing a copy of my paid receipt. When Seven Seas did not reply, it was clear that they had corrected their records and that the smudge on my name had been erased.

This was more than a lesson about saving receipts. It was much more. It was the discovery of the power of words. I was enthralled. Now I wanted to learn more words and find more ways to put them to powerful use.

All of this was happening in conjunction with Chief Lott's lessons about raising my head and speaking up, as well as Petty Officer Slikowski's admonitions that I bring more color into my life, along with my growing sense of self that came from my skills as a ship serviceman, and then the talking I was doing with Commander Williams. Out of all of that and my new love for language, I also discovered that I had things I wanted to say, opinions and arguments and thoughts and dreams and philosophies and stories that were boiling up inside me. My ideas brought forth questions about every aspect of human life conceivable, including the existential issue of wondering how a person could go through so much tribulation in life and then simply die.

After suffering and overcoming through lessons and loss, were we all destined to be mere dust in the wind? Or was there more? Was there a way to share what you've learned, a form in which to pass it on to others for them to use, from which to benefit and perhaps avoid your pain? Because, as far as I was concerned, the failure to find that form was to miss the boat entirely. What was the use of life, otherwise?

Walking tall into manhood, I now recognized that life should have purpose and a context in which to communicate that purpose to others. Like a life-insurance policy without beneficiaries, if there's nobody there to learn what you've learned, then your life is without consequence, no matter how much wealth of wisdom you've attained.

Writing, though I didn't know it yet, was to be my form. In the meantime, I had a hunger to benefit from the wisdom of others and thus I embarked on a learning kick. I read whatever I could get my hands on—everything of black interest; novels by James Baldwin; nonfiction books like *They Came Before Columbus* and *The Slave Community,* which Commander Williams gave me; a book that helped me explore my questions and skepticism about religion called *The World's Sixteen Crucified Sav-*

iors; and novels like *Crime and Punishment* that took me into places and times totally foreign from my own.

Reading wasn't always easy. In having me do some writing and reading out loud for him, Commander Williams diagnosed me as dyslexic. After all my years of struggling to read and hating to read out loud, I now understood that it wasn't because I was slow or less intelligent. I also understood that it was all right to take my time with books and their words. The effort was worth it, because in a book lay a whole other world to which I could travel, all while I was actually traveling on ships to other worlds within the world at large.

The more I read, the more turned on to language and its possibilities I became. I tried using different words in conversations with my shipmates, and when I didn't know how to begin a conversation, I would try and raise a historical topic and offer to debate that. At first, my efforts fell flat. But over time I actually developed some interesting philosophies. Through a more educated background on slavery, for example, I gained a different perspective on the Picketts and how the oppression they had suffered had perhaps turned them into oppressors. I became educated about black history that began not in slavery—as it was taught to us in school—but in our proud African heritage. From that, I developed what I called my toolbox theory. As I understood it, one of the many horrendous violations that slavery wrought was robbing black people of their natural language—their toolbox—and replacing it with that of the slave master. That meant those of African descent who grew up speaking English were being forced to use somebody else's toolbox. A person's natural language, I concluded, is the electricity of his or her soul, and to disconnect it is to shut them down.

312

The more I delved into the complexities and quirks of language, the more I appreciated the power and magic of words. It may be hard to conceive that all of this had sprung from writing a letter of complaint to Seven Seas. But it did. And then I took it a step further, thanks to a record by none other than the Dramatics. It was a beautiful nautical song with haunting lyrics: *"I'm the captain of my life, the master of my destiny . . ."*

So taken with the words, I wrote them out on the back binder of a photo album of mine. And when I wrote them out, I saw for the first time that the words rhymed. I wanted to make music with words that way, too, and I attempted to write a song without music, using part of what they'd written. Then I decided that I would write a song, without music, that was of my own creation. Not even having the term for it yet, I had become a poet.

On the U.S.S. *Cleveland,* I showed one of my shipmates my new creation. We were preparing to go overseas and he was already upset about being separated from the wife he adored. After he read my poem, he asked, "Would you mind if I gave this to my wife? She'd really love it."

Unprepared for his request, I thought fast and then said, "You can give it to her, but you have to buy it."

We had a deal. He paid me three dollars for my first poem and, true to his prediction, his wife was thrilled by it. The whole time we were overseas, he bought more poems. Word spread. Other guys started buying my poems to send to their wives and girls. Later, my reputation as a writer of romantic poetry swelled to such an extent that the chief storekeeper on the U.S.S. *St. Louis* dubbed me Sweet Mouth. By then, I had my own section in the ship's paper, with a monthly column that included my latest poem.

My navy reports from this time were almost effusive:

SH2 Fisher is always impressive in uniform. He has an excellent written and oral command of the English language. He fully supports the Navy Equal Opportunity programs. Petty Officer Fisher has tremendous potential and excellent leadership qualities. . . .

Even my written training reports received special mention by my captain for their excellence. It didn't occur to me yet that writing was something I would do beyond these contexts. All I knew was that I loved words and what I could do with them—to make people think and feel and see.

Words were my paints, thoughts my palette, paper my canvas, and the world itself my ever-changing subject. In the course of my eleven years in the navy, I got to know and see the world quite well. Twice I crossed the equator at sea. The first time it happened I became a shellback—an event so momentous, the navy had an ornate, archaic ritual for all of us pollywogs to perform. It involved swimming through wet garbage and crawling around the ship on all fours with a raw fish in each of our mouths as the shellbacks slapped us on our rear ends with wooden paddles and shillelaghs made from fire hoses. The next time I crossed the equator I was a shellback and was asked to administer the ritual to the new initiates. I refused. I understood this was old maritime deep-sea stuff, but I didn't like it.

On one of my first journeys we sailed north to Monterey and then on to San Francisco, where the sight of the

Golden Gate Bridge and the experience of passing under its magnificent shadow took my breath away. From our own continent down to South America, I sailed to most of the others—Africa, Asia, Europe, Australia. Destinations to which I traveled, some of them frequently, included the islands of the South Pacific, Saipan, Japan, Tinian, Guam, Goat Island, the Marianas, New Caledonia, New Zealand, Tasmania, New Guinea, Korea, Hong Kong, the Philippines, Thailand, Spain, and Mexico, to name only some. During the Iranian hostage crisis, we were in the Gulf of Oman for several weeks; when the confrontation with Libya and Gadhafi took place, we were in the Indian Ocean for months.

I had grown to feel at home in the world and on the water. If I had a favorite place, perhaps it would be in the middle of the Pacific at midnight, with dolphins playing and whales spouting and, in the clearest of skies, the stars so brilliant that it looked like morning. That was peace— the easy winds, the green glow of the water as the ship cut through it, and the salt always on my face.

Even when the sea was angry, I felt at one with it, knowing my own storms, my own upsets. I have a vivid memory of the time our ship's condensation machine broke. We were rationing water, a week away from our destination, and everyone was badly in need of showers. We sailed into a tropical storm and everyone went out on deck to bathe in the thundershowers. What a sight— everybody buck naked on the main deck. I stood just beneath the small boat rack where the water was collecting and draining off, creative as always. Not even shy without any clothes on, I really had come a long way from where I began.

For a kid whose whole world was once confined to the boundaries of the Glenville area of Cleveland, I learned my way around the map of planet Earth very well. Arriving in distant ports, I usually knew exactly where to go. Like New Guinea, where there's only one place to hang out in the whole country and I knew where it was, what road to take to get there, and when I arrived they knew me, too. By now I was often recognized on the streets of foreign lands. In Hong Kong at 747, a restaurant I frequented, they always greeted me, "Ah, Mista Feeesh, nice to see you!" Shopkeepers in Australia and Guam and Korea would lean out of their stores when they saw me coming, calling, "Fish, how are you?" urging me to linger for food or drink and conversation, or to buy from the shop.

Of course, out of everywhere that I traveled, I became best known in the seaside city of Sasebo, where I was based for two years. A small town in Japan's Nagasaki prefecture on the island of Kyushu, Sasebo is located at the southernmost end of the country and is the city where the bombing of Pearl Harbor was planned. The city was modern in many respects, but still old-fashioned in others—with signs of its antiquated sewer system visible, though no longer in use. Many of the older buildings and houses were built of teak and on hot summer days a teak smell, like a fragrant, subtle Oriental incense, filled the air as I walked through those older neighborhoods. Sasebo was clean, quiet, and so safe I once saw a woman out walking alone at one-thirty in the morning with her handbag on her arm. She was a block away from me when she passed by and as I watched her vanish into the distance, I thought, She couldn't do that in Glenville.

Sasebo was to me the kind of far-off, enchanted place I had read about as a child. It was a storybook place that seemed tailor-made for a romantic like me—twenty-five years old and a veteran writer of love poems who had never been in a real love relationship before. Sasebo changed all that. It was there, in my second year, that I first fell genuinely in love and made love for the first time.

On a balmy spring day, the sky a baby blue, all of nature's colors in blossom on tree branches and gardens, I left the base to do some shopping at a department store in the Ginza, the shopping district. Though I didn't leave the base very often, when I did venture into town the local people always seemed to notice me. To some, I was probably a horrible intrusion, to others a rare treat. The latter may have been the case at the department store where I realized a beautiful girl was watching me closely. This wasn't uncommon. Without being obvious about their curiosity, the younger Japanese people seemed to be fascinated by foreigners, particularly African-Americans.

What was uncommon was her striking beauty. She was tall, with beautiful Oriental features. Her hair, cut short, was dyed a reddish amber color, and she was wearing a spring-colored summer dress that reminded me of something I'd seen Doris Day wearing in the old fifties movies. Not only had this pretty girl noticed me, but she also was tailing me. In my younger days, I was always the one on the hunt, and this turnabout made me feel good.

I moved over to her and said, "Hello."

She didn't run, but responded, saying, "Hello," too, in a genuinely shy manner.

I was touched by her shyness, thinking how that had always been my role. We introduced ourselves to each other. Her name was Seiko Fukashima and she was in Sasebo, where she grew up, to visit her mother. Seiko was living at the time in Tokyo, working as a model for the cosmetics company Shiseido. Surprisingly, we spoke enough of each other's language to communicate and we found ourselves setting off for a coffeehouse where we talked cryptically for hours.

For days and months to come, Seiko and I met at that same coffeehouse whenever she was in town to see her mother. Then she began to make special trips to Sasebo to see me. We became close and we both fell in love.

After so many years of painting poetic images in my mind of the ultimate romantic evening, I experienced the real thing when I went to visit her in Tokyo, where I checked into the Hill Top Hotel. Seiko took me around the city to show me some of the sights and after a long day out together, we returned to the Hill Top and went, hand in hand, to my room. For many long minutes, we sat together on the edge of the bed, neither one of us speaking at all. Sensitive to my mounting embarrassment, she stood, turned off the lamp, and began to undress.

A soft glow from a streetlight outside the window made its way past the blinds and flickered like a candle in the dimly lit room. I could see the outline of her tall, sleek body as she came toward me, amorously, dressed in her bra and panties.

Moving in close, she took my head in her hands, gently pulling me to her bare midriff. Still seated, I wrapped my arms around her waist. We came together inch by inch, still slow, savoring every touch and feel of each other. Soon she knelt down and began to undress me, helping me off with my shirt and pants. At the moment that we were both completely naked, my old fears flamed from the caves of the past. But when I looked in her eyes I saw only a reason to trust, and my fear was extinguished.

Wrapped around each other, we kissed and caressed, softly, gently, until we made love in a way that was caring, tender, beautiful, and emotional.

318

Since Seiko was more experienced than I was, she took on the role of teacher in this area. But, at twenty-five years old, I was an avid student, eager and ready to learn and practice all that I could—again, making up for lost time.

The rest of the year unfolded in a blissful blur until we both realized that the time for me to return to the States was drawing near. With each passing month, the lump in my throat grew. I had to go; I had by this point what I called a "two-year" rule, having established a pattern not to stay on any ship or in any position longer than two years. There was never a discussion about whether we would try to maintain a long-distance relationship; we had an unspoken understanding that she had her career ahead of her, while my naval duty and journey into manhood were elsewhere.

On the scheduled morning of my departure, Seiko appeared at the main gate of the base in Sasebo and surprised me by saying that she had arranged to accompany me to Narita airport in Tokyo, where I was to catch the flight that would take me out of the county.

We traveled by bus to the Nagasaki airport and boarded a flight to Osaka. Filled with many emotions, I was feeling very self-conscious. On the airplane, Seiko noticed my discomfort and asked me what was wrong.

"People are staring at me."

"They stare at you because you are handsome, Antwone." She smiled, as if to say, Don't you know that?

Wow . . . I'm handsome. I didn't know it, not because the person I saw in the mirror was unattractive, but because no one had ever said that to me in my life.

After a few hours' layover in Osaka, we took another plane to Narita and went to wait in the gate area. In real time, it had been a long day, although it seemed to have sped by, and now the minutes fled past like seconds until the final boarding call for my flight was made over the loudspeaker.

We stood up from our seats and she walked with me as far as she was permitted. We kissed a sweet kiss and said good-bye. As I made my way to customs, walking along a glass partition, I could feel her eyes on me, so I paused and turned to look. There she was, crying, watching me go. The sight burst the lump in my throat that had been there for a month and the liquid rose and my eyes welled up with the knowledge that I would never see her again.

When I reached Guam, I discovered that Seiko had placed a bottle of her perfume in my sea bag. Her scent stayed with me, as did her memory, for years.

When the day came for me to say my final good-bye to the navy, I was ready. I was proud of the man I had become and of my accomplishments, including those detailed in my record from my last years of service:

```
SH2 Fisher is one of the finest bar-
bers in the Navy. . . . His military
appearance is always neat and his
behavior is excellent. During a com-
mand    inspection,    Petty    Officer
Fisher was one of three to be recog-
nized by Rear Admiral Butcher for an
outstanding uniform. He gets along
well    with    all    members    of    the
crew. . . . Responsible for the day
to day operation and supervision of
laundry operations, barber shop, and
damage control . . . accountable for
```

$15,000 inventory of ship store items. . . . As services supervisor he has ensured that uninterrupted laundry and barber shop services have been provided to 325 crew members and 315 embarked USMC personnel . . . despite numerous problems caused by outdated equipment and an undermanned division. . . . Awarded battle efficiency "E" Ribbon. . . . His leadership and supervisory skills enabled to experience ZERO discrepancies during inspections . . . faced with severe shortage of manpower, he has reacted with a "can-do" attitude . . . Made arrangements for the delivery of flowers and gifts for crew members during Thanksgiving, Christmas, Valentines and Easter. . . . Spent off duty hours participating in orphanage rehabilitation projects in Korea and Hong Kong.

It made me very proud to be relied upon to handle thousands of dollars of goods and cash, ethically and responsibly. Considering that I never had family to remember on holidays, I was proud that I took the initiative to make sure the entire crew sent gifts for holidays. Volunteering at the orphanages was one of the most personally rewarding things I could have done—seeing the children who reminded me so much of myself. The medals and awards I received made me proud: two Good Conduct medals, a Sea Service Deployment ribbon with three bronze stars, and a Humanitarian Service medal for taking part in operations to save boat people fleeing from persecution off the coast of Thailand.

At the end of more than a decade, I was grateful for everything I received from the U.S. Navy. At first I had enlisted because I needed a safe place to be, because there wasn't anywhere else for me. The navy was my haven. But after a time, it became much more: It was family, friendship, belonging, education, and, ultimately, purpose. Not only was it important to know I had something to offer, but it was important that I was doing it on behalf of the United States of America. I wore my uniform with pride.

At almost thirty years old, I felt I had also given back for all that the navy had done for me, and I made plans to enter civilian life. Rather than face drastic culture shock, I sought an opportunity to go to Georgia, where I would be trained as a federal correctional officer and then begin work at the federal correctional facility at Terminal Island, off the coast of California near Long Beach. Still a child of institutions, I would be continuing as an employee of the U.S. government.

In my final months of naval service, while stationed at the navy's Long Beach Memorial Hospital, I took care of a long-needed surgery to repair a congenital problem in my jaw that was now creating other health concerns. My good-natured, middle-aged surgeon, Captain Anderson, prepared me by saying, "This is by no stretch a minor surgery." He wanted to make sure that I listed a next of kin in the event that something went wrong.

Eleven years earlier, when I enlisted in the navy, the recruiting officers required that I give them an address. At that time, the only person whose address I knew to give was Mercy. This was different. The truth was that I had no next of kin. But because I had to give them a name in order to have the surgery, I made one up.

The notion that I knew no one to whom I was blood-related began to bother me. That, together with the simple statement Commander Williams had made to me some years before—"Have you considered that you come from somewhere?"—was to become an increasingly urgent preoccupation. It seemed I had traveled to every far-off shore except one: the place where I came from, wherever it was.

As I write these words, another ten years have gone by since I left the navy. And yet there is little I've forgotten. How could I? I've spent hundreds of days of every season at sea, sailing sea after sea after sea, and I can tell you that there are more than seven. I became one of a million or more sailors who've passed through time, who can boast of requited love relationships with the sea, of days of clear blue skies and nights where the stars alone light the sky. To experience sunrise and sunset at sea is to experience the full beauty of God's imagination.

I discovered that every ocean has its own unique personality, that the Pacific is welcoming, the Atlantic broods, the Antarctic is as silent as the moon, and the Indian Ocean, with its atmosphere of mystery, burns like the Sahara.

My eyes have taken billions of photographs of the countless scenes of my travels, travels that have taken me so far away from my beginnings and so far beyond my dreams. The image of my face is locked in the memories of many and theirs are locked in mine.

I'm beached now, never to see the world by sea again. But I know the seas will not forget me, nor the eleven years we spent together.

eight

The decision comes to me one night in the summer of 1992 in my single-room apartment in Los Angeles, as I'm stretched out on my bed, daydreaming—still my favorite vice. It's provoked by the vague discomfort I've been feeling for over three years, ever since Captain Anderson raised the issue about my next of kin. Not knowing where to begin searching, I've usually managed to put it out of my mind. But not tonight. Tonight it comes at me with new urgency, as if I'm hearing whispers from ancient African ancestors, continents and lifetimes away, inviting me to come, at last, and find them.

They're inside of me, I know, like the pancake dream I cherished in my childhood, just beyond the barn door. But where's the road that leads to the barn? Where's the grassy field? Who is that strong man and that kind woman to show me the way?

Just then, as I ask these questions in my mind, I have an idea. It starts like a tiny light in the distance, a firefly moving toward me, barely visible. It begins with the image of Ms. Nees, the caring young social worker I met for the first time that stormy day I left the Picketts' house. I can see her head bent over my file and then raising it up as she read from it and spoke to me, as though offering me clues to myself. I can see my younger self, sixteen years old, sitting next to her with the drone of rain in my ears, watching the downpour outside the window next to her desk.

Tonight, the sights and sounds out my open apartment window are so different from that terrible day in Cleveland. It's a balmy L.A. night in July, almost time for my thirty-third birthday. Palm trees that line the boulevard sway in the wind like they're waving at me, lights streak across my walls from cars cruising the neighborhood, and the strains of hip-hop and rap waft up from other apartments and passing cars.

My apartment is small, simple, and scrubbed clean to a shine. After all, I was in the navy for eleven years. With echoes of SH1 Slikowski in my ears, I've remembered to put some color into it, too—hints of gold, rich browns, even a touch of burgundy.

All in all, my life isn't bad. I have friends. We go out dancing and drinking—something I learned how to do in the navy—and I date from time to time. Nothing serious. With my two-year rule, I avoid long-term commitment, feeling that I want to make something of myself first. But what that something is, I'm still not sure.

For the time being, I've just started a new job as a security guard at Sony Pictures Entertainment, yet another whole new world to which my ship has sailed. It's not what I want to do with the rest of my life, but it's a welcome change after my three years as a federal correctional officer. One of the biggest shocks is looking at my paycheck, the first I've ever received that doesn't have the Statue of Liberty on it. Another ironic switch is that at Terminal Island I was a guard hired to keep people inside from busting out and here at Sony Pictures I'm a guard trying to keep people on the outside from breaking in.

Of course, that's not my official job, but in my short time being in charge of patrolling the studio lot, I've observed that show business operates very much as an exclusionary club. Then again, I wonder sometimes if the people on the inside are actually happy, if maybe they're not prisoners, in a way, of their need to be important.

That seems to be the prevailing obsession. Not everybody comes right out and says it, but some have: "Don't you know who I am?" The people with attitude who say that or want special treatment aren't the actual power players. It's usually the assistant to the producer, or the brother of the director, or the friend of the star's cousin. Meanwhile, the truly successful and powerful I've met on the lot are often the nicest—such as Danny DeVito, Tom Hanks, Dustin Hoffman, and Rob Reiner, who were each warm and friendly, even to the point of remembering my name. A "How ya doin', Fish?" from a busy, important person, or from anyone, for that matter, can always make my day.

Seeing famous people come and go has become old hat. But not the day I saw Sidney Poitier. I almost knelt down. He reminded me of a prince, a legend, and still down-to-earth enough to smile and say hello.

Aside from the less appealing exclusionary aspects of show business, I can also say that after a few weeks on the job I feel an energy that dominates the place, an air of excitement and hope. Everybody's hoping their current project hits, their last bomb gets forgotten, their next idea gets bought. In fact, with so many screenplays lying around in the different offices I inspect, I've thumbed through some of them and started to read passages. Like learning a new language, reading scripts takes practice, but I'm keeping at it. I suppose it's my way to better understand this world where I happen to be working for the moment.

One thing I've discovered that the movie studio and the prison populations have in common is that they're both full of characters. At the Federal Correctional Facility at Terminal Island, where they came from every walk of life, I became a virtual connoisseur of characters. Some of them were famous, too. Dr. Jeffrey MacDonald, convicted of killing his wife and children, was there, a respectable-looking man who spent most of his time in his cell and the law library, when he was allowed, working on his appeal process. The well-known white-collar criminal Barry Menko, the Z-Best carpet-cleaning guy, was there, too. Peter Milano, a nice older gentleman, reported to be in organized crime, told me there was no such thing as the Mafia. Jack Battaglia seemed like a regular, nice, working-class guy, in spite of his alleged Mafia connections. Mario Gambino was from a New Jersey crime family, another good fellow who was arrested for his involvement in a drug ring that had pizzerias delivering drugs in pizza boxes.

When I started at Terminal Island most of the inmates were there because of white-collar crimes. That changed when President Bush launched his War on Drugs and established the federal crimes and sentencing guidelines. Then we started getting gang members, from black gangs like the Crips and the Bloods, the Latino gangs, the white supremacists, all the different groups, bringing with them many of their street business practices.

Mackey Boy, a Crip, had a few deals going, including providing protection for certain inmates like this older Jewish man. Mackey Boy was also pimping homosexuals on the yard for quarters, since that was the largest money denomination allowed in the facility. One of Mackey Boy's homosexuals was Rice, who had breasts. Rice had started his sex change before getting caught in a bank heist scam where he was working and trying to steal money for a full sex change.

As much of a bad-ass as Mackey Boy was, I once saw him shaking in his shoes when Peter Milano cornered him and talked to him in the most menacing voice, saying, "Don't you know I will bury you?" That was all that was said, but I'm sure Mackey Boy never caused any trouble around Peter Milano after that.

As a kid, one of the ways I learned to cope with being afraid all the time was to scare somebody else. Dealing with the inmates in the correctional facility, that skill was worthless since most of them didn't scare. So, as a federal officer, my approach was to treat everyone the same, giving that person, regardless of his crime, his dignity. I had the reputation for being fair and level-headed. One day, Jack Battaglia looked up to see me heading his way and he announced to all in earshot, "Now, here comes a gentleman."

If being a gentleman in his eyes meant recognizing everyone's humanity, then I accepted that I was a gentleman.

These characters were all great storytellers. Most of them would usually tell you that they didn't do whatever they were in for. It was a policy of mine not to ask what they were in for, even though I became an excellent sleuth in such things. Just by meeting someone and sizing him up, I could almost always guess exactly what kind of crime he had committed—or hadn't. Whether they were bullshitting or not, I enjoyed the stories and the life philosophies they'd learned in spite of their falls. Mario Gambino once said to me, in his thick Sicilian accent, "You know, Fish, opportunity comes in life only a few times. And when it does, you gotta grab it." To underscore the importance, his right hand made a gesture of reaching out and grabbing something supervaluable from out of the air.

At first, I thought he wanted me to bring contraband into the prison. But when he didn't ask, I realized he was just offering solid advice.

The decision to leave my job at Terminal Island arose after a difficult incident during which I came upon an officer beating up an inmate in the SHU, the special housing unit, and stopped the beating. The officer was later dismissed and my immediate failure to report it resulted in a ten-day suspension. In spite of my suspension and my silence, I was labeled a snitch by some of the other officers who wanted to hold me responsible for the officer's dismissal. The stress started taking a toll, and when I cursed out an inmate one day, I knew it was over. Plus, I had well overstayed my two-year-limit rule. After the job at Sony presented itself quickly, I said good-bye to my last government institution and entered the real civilian working world.

What I learned more than anything from my three years at Terminal Island was how easy it can be—for anyone—to get into trouble and wind up there. I hurt for the tragic waste of human potential I witnessed at Terminal Island. So many faces there were reminiscent of others I had known earlier in my life and had passed by along the way. Many were the Dwights of the world, naturally brilliant, naturally endowed to make something great of themselves, lacking only in love to fuel them perhaps. Statistically, given the sum of my circumstances, I was probably the most likely candidate to have ended up behind bars. But having defied statistics, my last institutional port showed me the powerful stuff of which I was made. It was a lesson about the grace of God, about recognizing the inner wisdom that had directed me to create rules for myself early on; and it was a reminder of how fortunate I was that whenever I was really heading toward trouble, each time a guiding hand or an opportunity had steered me away.

In my apartment in L.A., as I'm starting to focus on the memory of Ms. Nees telling me things about myself and my history, the lesson I learned in the correctional facility about how easy it is to get into trouble makes me think of Eva Fisher. It's interesting, as it occurs to me now, that I went full circle—starting my life at birth in a correctional facility and returning in my adulthood as a guard and observer, as if returning as a witness to the scene of the crime to better understand her—my mother.

This is when the idea strikes. To begin a search for my next of kin, why not start there, with the first documentation of my existence in that very file that Ms. Nees was reading?

Over the next two days, I make a series of phone calls to Cleveland. I request a copy of the file maintained on me during the years that I was a ward of the state, and I am informed that because I was never adopted, I'm legally entitled to receive a copy.

On my birthday, I sit down to read the reports that tell the story of my first eighteen years of life, told from the point of view of thirteen social workers and two psychological examiners, each one of them, through each entry, corroborating and validating the truth of my memories.

Through my file and my contacts at the social services department, I begin making inquiries about the current whereabouts of Eva Fisher. Finding nothing, I start to focus on the man that my mother claimed was my father, Edward Elkins, whose name appears twice in my file—once in reference to Eva Fisher telling her social worker that he had been killed two months prior to my birth, another time when the name was raised by a social worker to Mizz Pickett, who said she heard from relatives that he was a thief.

Then I suddenly remember something Ms. Nees said that isn't in my file: "Your mother told her social worker that she heard there was an article about Edward Elkins's death that was published in the *Cleveland Call and Post*."

Without giving it a second thought, I reach for the telephone and dial information for the Cleveland Public Library. Early the next morning, before I'm due at Sony, I place a call to the library and inquire about the availability of articles from the *Cleveland Call and Post* going back to June 1959. A businesslike young woman at the library takes down the name *Edward Elkins* and promises she'll look in the archives for any article detailing his death. "Call back in a couple of weeks," she warns. "It might take some time to locate it."

Two weeks later, to the hour, I call back and the businesslike young woman reports, "Yes, Mr. Fisher, I found the article you were looking for. June 13, page 3A of the *Call and Post*." She explains that if I send her a money order, she'll mail me a copy of the article. Another week and a half go by and finally I am sitting in my apartment, opening up the envelope, taking out a copy of a copy of the microfiche article with a picture of my father that I can't make out. Everything else is legible, starting with the headline, *"YOUNG MOTHER FREED, Calls Slaying 'Justified'"*:

A willful young lover who friends believe "meant to marry the girl" was shot and killed Thursday noon as he attempted to force his way into the E. 90 St. bedroom of the girl who had borne him two children but was "tired of his abuse." The victim, Edward Elkins, 23, of 975 Parkwood Dr., was literally blasted down the stairs from the second-floor chain-locked door of Frances Holden, 19, who "couldn't take it no more."

For days, I sit with the article, reading it over and over, sorting out my confused feelings: about the man who was supposed to be my father; about Frances, the girl who killed him and deprived him of his life; and the two girls mentioned in the article, ages eighteen months and four months, who would be my half sisters. After a week goes by, still not sure how to act about this new information, I reread the last sentence for the umpteenth time:

Funeral services were tentatively set for Tuesday of this week at the Cummings Funeral Home on Cedar Ave.

Again, without stopping to think, I hurry to the phone and call information for Cummings Funeral Home. At first, I find no listing, but I try again, eventually locating the Cummings and Davis Funeral Home. A request for the death certificate and a check mailed off to have them send me a copy also yields me the address of Edward Elkins's next of kin, who signed the death certificate: 975 Parkwood.

From my long-ago days of walking the streets, pining for Freda, I know that very building, a four-family apartment building on Parkwood. It stands next door to Freda and Mona's house. Could it be that I was walking past the home of my relatives, all that time?

All of this is coming to me so easily, I figure it's worth more of a search and I order an Ohio Bell telephone book for Cleveland. But when it arrives, there are so many Elkinses listed in the greater Cleveland area, I don't know where to begin. Calling up strangers isn't my idea of how to spend a summer's evening, not to mention the fact that on my security guard salary, I don't have the money to pay for a costly phone bill. And so, what has seemed so far like a fruitful search comes to a sudden halt.

A few months pass. In late September, the smell of night-blooming jasmine thick in the cool evening air, I'm cleaning my apartment and move my bed to vacuum underneath it. Lying there on the floor, among a few other items I can't fit neatly anywhere else, is the Cleveland phone book. Again, I flip through the pages of listings for the different Elkinses, running my eyes up and down, imagining that one of the names might send me a special feeling. When none of them do, I decide that I'll allow myself only one call. Why not? Using the logic that the west side, as I remember it, is predominantly white, I concentrate on east side phone numbers, finally settling on an address not far from the Glenville area, which turns out to be the phone number of Annette Elkins.

The woman's voice at the other end of my one call is warm. "Hello?" she says.

"Hello," I say, nervously. "My name is Antwone Fisher. I'm calling long distance from Los Angeles . . . Uh, is this Annette Elkins?"

"Yes?" She answers in a way that tells me my name means nothing to her.

I continue. "I'm looking for the family of an Edward Elkins and I was wondering if you might have a relative by that name?"

"I have a brother by that name," she says. "But he's been dead a long time." She pauses, then asks, "Who is this?"

Softly, I say, "I . . . I think I'm his son." I pause, then say to her again, "My name is Antwone."

Another stretch of silence follows which she breaks by

saying, "Well, if you are Edward's son, you have a big family."

In a whirl of disbelief that my one and only phone call has allowed me to hit the jackpot, I tell Annette my birth date, my mother's name, and that I was raised as a ward of the state, mainly in a foster home in Glenville. Annette tells me that if anybody in the Elkins family had known about me, they would have found me and brought me home to be raised by them, my true family.

Holding back my tears, I listen as Annette tells me more about the family I've been missing my whole lifetime. After Eddie's death, the family soon left the house on Parkwood and moved to another one that was two blocks away from where I was growing up in the first house I lived in with the Picketts. She says that her parents, my grandparents, Horace, Sr., and Emma, died some years earlier. Still living in Cleveland are her twin sister, Jeanette, her sister Ann, and her brothers, Horace, Jr., and Spinoza. Her sister Eda, she tells me, lives in Chicago. "Oh, my goodness," says Annette, "and you have an uncle who lives in California."

"Really?"

"Yes, Raymond and his wife, Pat. Let's see . . ." She rustles in her address book. "They live near San Diego. Is that far from you?"

"Not too far." I'm smiling at her Cleveland notion of the world, where everywhere in California must be near to the other.

Annette gives me Raymond's phone number, takes mine from me, and then decides to give me the phone numbers of my other aunts and uncles.

My aunts and uncles. I repeat the phrase in my mind that I've never uttered before. I hang up already feeling all kinds of new and unidentifiable sensations. Thirty-three years old and in one phone call a gaping hole in my existence has been filled. I come from somewhere and I have found my next of kin.

Like new shoes that need wearing in, I allow a few days to go by so I can let my feelings settle a bit. On the evening that I'm planning to make some phone calls to my other relatives, the phone rings. It's Aunt Eda, my aunt from Chicago.

Within five minutes, I feel as if I've known her for years. Eda is articulate and well read, a lover of poetry and great writers like Shakespeare and Edgar Allan Poe, something she says she inherited from her father, Horace. "And your father, Eddie, was a wonderful writer. I'll never forget the letters and poems he sent home when he was in the army."

I tell Eda a little about my story, nothing too heavy, but about going into the navy and also writing poems. In fact, I offer, I could send her some.

When she receives the poems I've sent her, Eda immediately writes back, so touched and impressed with my talent, she says, declaring with amazement how much my writing reminds her of my father's. During our next phone call, she asks about the possibility of my visiting Cleveland so that the family can meet me in person.

"Well," I answer slowly, not wanting to admit that I can't afford a trip for the time being, but also not wanting to make anyone think I'm looking to get anything from them, "I'm new on my job and won't have earned any vacation for a while. I hope I can come next summer."

Though I hear disappointment in her voice, she assures me that everyone is happy about the prospect of meeting me and they'll just have to be patient.

In the meantime, I've been in touch with my other aunts and uncles, including Raymond in San Diego, who has invited me to drive down to San Diego to stay with him and his wife, Patricia, for the weekend. On a beautiful October Saturday, I make the two-and-a-half-hour drive, a straight shot down Interstate 405. Before arriving at Ray's house, I stop by the navy hospital, near the old stomping grounds, where there are shower facilities. Wanting to be clean and fresh when I meet my first relative, I shower and change and then drive to my uncle's.

Patricia greets me at the door, offering a warm smile. "I only saw your father once," she says, "and I'll never forget him. You are his spitting image."

Over her shoulder, I spot Raymond coming toward me. Tall, handsome, with intense flashing eyes, he looks like an older me. This being the first time I've ever met another human who looks anything like me, I stare right at him. Saying nothing, he stares right back, as if seeing a ghost. Finally, he breaks the silence with a big grin and locks me into a bear hug of an embrace. When he releases me, I turn to gaze around his home, at the bright array of paintings, like an art gallery. They're wild and brilliant. Raymond, it turns out, is an artist, a painter. And something of a musician, too.

Over the next several hours, Uncle Ray regales me with stories from the Elkins family past. Mixed in between bits of family history are other, taller tales, more improbable but colorful excerpts that hold me at the edge of my seat. Patricia interjects now and then, telling me that she lived for many years a few doors down from where I'd lived with the Picketts. In his youth, Ray says, he was a neighborhood brawler. "They used to call me 'Parkwood Bruno.'"

As the puzzle pieces begin to fall into place, we come to the joint conclusion that I must have walked past my grandfather countless times as he sat on his porch, keeping an eye on the block.

"Yep," Ray says, "you look like your granddaddy, same Choctaw nose, same build."

When I tell about playing in the area of Parkwood where they lived, sitting on top of the mailbox and teasing the policeman, Raymond squints his eyes. "Man, I seem to remember seeing you up there. I can just picture a little boy up there. All the time."

"Well, I was a fixture on that mailbox."

They were so close, all that time. Just beyond the barn door.

Ray wants me to know how highly the Elkinses were thought of—intellectuals, artists, readers. The Bible and the collected works of Shakespeare were the two most prized books in the home. Horace, Sr., a medical doctor, was known as the most brilliant black man around, Ray asserts. "But he was no-nonsense, too."

He tells me Spinoza's favorite story about the time when Horace, Jr., the oldest brother, was the Minister of Defense of the Black Panthers in Cleveland during the Vietnam protest era. Horace, Sr., being of the old school, was unsure of the way the young people were going about forcing a change, and when he found his son in his leather coat and beret at home at the dinner table, he asked, "What are you doing here?"

"I came home to eat," said the younger Horace.

"Ain't you the head Negro in charge of the rest of them Negroes out there?"

"Yeah . . ."

"Then go eat with them!"

We laugh heartily. As Ray talks about Eddie, his singing talents, his temper, the way the girls used to call him "Swami" because he hypnotized them into doing what he wanted, I can feel the sadness of the family's loss. Something about my father makes me think of my friend Jessie—the boldness, the fearlessness, the not caring what anybody else said or thought until it was too late.

The subject of Eddie's death turns Ray somber. It is here that I begin to sense the family taboo against talking about it that has been in place for thirty-three years. The one thing Patricia does tell me is that she always remembered a nurse that day her mother came over and told Emma about the handsome boy who had been brought in to the hospital, practically severed in half, not knowing that it was Emma's own son.

In my mind's eye, I'm seeing the hospital they're talking about—the very same hospital that I'd always been so compelled to stare at, for reasons I never understood until now.

Later, I'll hear more from other family members—about Eda and Ann trying to offer Eddie help to get him away from the trouble, about Emma's premonitions on the morning of the shooting, about Horace, Sr., breaking down at the funeral and crying like a baby, and about the oldest of Eddie's two daughters by Frances remembering what he said to her mother after she shot him: "You killed me."

That evening, Ray takes me out to his favorite hangout where everyone knows him and he proudly presents me as his long-lost nephew. Accompanied by the band performing that night, he stands and dedicates his first song to me and croons in the sweet falsetto his brother Eddie taught him how to sing. Uncle Raymond's song of choice in my honor: "Daddy's Home."

* * *

Two days before Thanksgiving, a month after meeting Raymond, I made my fateful trip back to Cleveland. It had been arranged that I would fly to Chicago's O'Hare Airport, where I would meet Aunt Eda and the two of us would continue together to Cleveland.

Everything had come about very much by surprise when Eda called and invited me to come for the holiday, saying everyone in the family wanted to contribute to pay for my ticket. She never alluded to whether they knew it was financially difficult for me to pay, only that she hoped I could get permission from my work to let me take the time off.

When I went to my supervisor at Sony, I found it necessary to tell him my story to show that this wasn't a frivolous request. Not only did he give me the time off to make the trip home, but also unknown to me then, he began telling some of the studio executives about my story. Little did I know how pivotal that would be in my future career.

My trip lasted ten days. Each moment, each encounter was more poignant and happier than the last. Determined to maintain composure, I felt as though I had a protective shield around me the whole time. Like Raymond and Pat, my other aunts and uncles couldn't stop saying how much I looked like my father. Annette and Jeanette, the twins, pointed out that I moved like Edward, too—like a tiger, they said. At Spinoza's, where I stayed most of the time I was in Cleveland, he and Horace, Jr., Eda, and Ann remembered that my father, like me, had a habit of talking under his breath when he was walking away from people, with sort of a sarcastic sense of humor.

All of the Elkinses I met were attractive and tall, all with unusual-colored eyes—greens, silvers, and browns. That Thanksgiving was majestic. There were more relatives to meet—the spouses of my father's siblings, my cousins and their families. The love and happiness I felt coming toward me and flowing out of me was like a powerful cleansing, like taking a long, long shower after being gone too long in the wilderness. It was the scene of my dream. I was the guest of honor. Only instead of eating pancakes, we were eating turkey and gravy, stuffing, cranberries, sweet potatoes, and all kinds of cakes, pies, and desserts. Uncle Horace had a quality that reminded me of the tall, muscular laughing man from my dream, and Aunt Eda was the grand loving woman who whispered in my ear to give me courage.

I was given gifts of photographs of Horace, Sr., Emma, and, of course, Eddie. There he was, my father, in uniform at all of nineteen years, looking ready to take on the world, with a smile that could light up the night sky. Before I left Cleveland, I went to the gates of Calvary Cemetery, where Edward Elkins had been buried two months before my birth. Because his grave was marked only by a number, I did not attempt to locate it. In time, however, I would visit that spot and later arrange to have a headstone placed there to commemorate him.

At a certain point in all the festivities, Spinoza got it in his mind that he was going to locate Eva Fisher. He took me out to lunch, promising me the best fried fish in Cleveland, then around to various different joints he knew to ask around to see if anyone had heard of Eva Fisher. At one place—what looked to be an abandoned house where people were gambling with stacks of money—Spi described the age of the woman we were looking for and somebody said, "That sounds like Mae-Mae."

But whoever Mae-Mae was, the person didn't know where to find her.

Early in my visit, my half sisters, Renee, two years older than me, and Pamela, a few months older, came over to Spinoza's to meet me. There was no mistaking our family resemblance. To present myself as nicely as I could, I had brought along my best clothes, and my sisters complimented my good taste. I was wearing two gold chains around my neck and took them off, giving one to each of them.

Pamela gave a party that night in my honor at her and her husband's house. The party hadn't been going on long when I noticed a woman in her fifties staring at me. I was getting used to people reacting as if seeing a ghost when they saw me. But her expression was filled with such fear and discomfort, I knew she had to be Frances. Wherever I moved in the party, she followed me with her eyes, averting them as soon as I looked back toward her. She was obviously still in pain and I felt that maybe she was feeling guilty or worrying that I blamed her. In a quiet moment, I went and knelt down next to where she was seated and took her hand, patting it gently. Her hand was soft and fleshy. "I'm Antwone," I said.

"Hi, Antwone," she said.

When she continued to stare at me, I felt uncomfortable and stood up, nodded good-bye in Japanese fashion, and walked away.

I slept that night at Pamela's and the following morning both Renee and Pamela came in and sat on my bed and woke me up.

Opening my eyes in surprise to see them there, I thought, Well, this must be what sisters do.

"Spi called . . ." Pamela began.

Renee interjected, "He thinks they found your mother."

"Do you have any Jack Daniel's?" I asked Pam. She gave me a look like I was kidding. When she saw I

341

wasn't, she brought me a bottle. After getting dressed—in a long black dressy trench coat almost to the ground, over silk shirt and pants, burgundy alligator shoes, with my black sunglasses—I took a long swig off the bottle.

By then, Spi had arrived to pick me up. In the car, he recounted how his wife had the idea to call a longtime family friend, Jess Fisher. Apparently, Jess was a take-care-of-business, down-to-earth guy, retired from construction work but still a dues-paying union man. Spi thought it was a good idea, since he did have the same last name as me. "So I called him," Spi said, "and asked him if he knew anybody named Eva Fisher."

"Yeah?"

"Yeah. So Jess says, 'I gotta sister Eva Mae Fisher. They call her Mae-Mae.'"

We went to meet Jess Fisher, a short, strong-looking, brown-skinned man, who was as to-the-point and no-fuss as Spi had said. Meeting the first relative on my mother's side, I saw similarities between him and me that were different from the Elkins family traits I'd been noticing.

"I'm glad to meet you," I said.

"You can call me Jess. Or, you can call me Fish." He grinned.

"You can call me Fish." I grinned back.

I showed him my birth certificate, and he noticed that the address used as my mother's permanent address was his address in 1959. Then Jess announced, "Okay, let's ride."

When Jess went to get his cap, I looked at Spi and said, "You aren't gonna let anything happen to me?"

Spi seemed confused by my question and then relaxed with understanding. "No, man," he promised, "I'm not gonna let nothin' happen to you."

Jess parked his car in front of a Longwood housing

project. We walked up the walkway, me following my two uncles, making our way down to a lower apartment entryway.

The door was answered by a girl of about twenty-three. She had features similar to those of Jess, and the two obviously knew each other.

"Hey, Melody, how you doin'?" Jess said.

"Oh, hi, Jess," she answered, and turned for us to follow her inside.

The door opened into a long, narrow kitchen area. My heart was beating rapidly. I had prepared a script in my mind for what I needed to say to my mother that had been painfully written over a lifetime. I would ask her: Why didn't you ever come to get me? I would ask her, Didn't you wonder about me? What I was doing? What I had become, or even if I was still alive?

She would have to hear me say, I dreamed about you every day, my mother, what you looked like, your voice, even your scent. For thirty-three years, I've dreamed of you. Didn't you miss me at all?

I would let her know that I'd taken care of myself all my life, that I'd never been in trouble with the law, that I'd never fathered children, and never done drugs or smoked a cigarette in my life.

I've educated myself, I'd say. I've read hundreds of books. I've traveled throughout the world. I speak two languages. I've served in the U.S. Navy and been awarded medals and ribbons of honor. I've been trusted with people's lives. I paint and write poetry. I have friends who would help me if I needed help. I made my way through terrible times, and I never complained. I've become a good man and a good person.

All of that I had prepared to say. But when I walked into the dimly lit apartment with its shabby furniture and damp smell, I became unsure. And the moment that Jess asked Melody, "Where's Eva?" and Melody said, "She's here," and I turned to see a frail woman in nightclothes leaning into the refrigerator, my reservoir of hurt and anger emptied out full.

My dark sunglasses hid my shock at seeing her, a woman who looked much too old to be my mother. She was wearing masculine glasses and her thin hair was uncombed. Aside from being dressed in her nightgown, she appeared to have no teeth.

Jess later told me he didn't give any of us any advance warning on purpose, so everybody could see what was what, he said, the real deal. Intent on making sure there was no hiding, Spi saw me standing there in my sunglasses and said, "Take those off."

Jess greeted his sister, "How you doin', Eva?"

She shrugged. "How you doin,' Jess?" Her voice was heavy, rough, and raspy.

He didn't answer. Instead he stepped aside, so she could get a full view of me. "Who is that?" Jess asked her, pointing toward me.

"That's Johnny."

"No, that's Antwone Quenton Fisher," said Jess.

"Who is Antwone?" she began to ask, but before anyone needed to remind her, she made the connection and started to moan, losing her footing and holding on to a chair, "Ooh, God, no . . . ooh, my God! Oh, Lord! I saw ya'll get out de car and I thought it look like Eddie . . . Oh, my God, please no!" She moaned and wailed for several minutes.

When my mother began to quiet, Jess said, "Everybody sit down." Everybody sat down at the kitchen table, Eva, Jess, and Spi. Melody was still standing. "Have a seat?" Jess asked me.

"I don't wanna sit down," I said, though I was still reeling from shock.

My mother stood up just then and walked toward Spi, looked at him, but spoke as if to me, "Yuh wont som't'eat?"

Jess and Spinoza waited for me to respond.

"You talking to me?" I asked.

She continued to look right at Spi as she rolled her eyes. As she turned to walk back to the stove, she made that gesture at me that kids do where she circled her ear with her finger, gesturing me to be crazy.

I hadn't seen anyone do that since I was a kid. I sat down with the others and watched her stirring something in a pot. For several minutes, a tense silence filled the kitchen, broken by periodic clanks of her spoon against the pot.

In the place inside me where the hurt of abandonment had been, now only compassion lived. The speech I had prepared was meaningless. What I understood now was that even though my road had been long and hard, my mother's had been longer and harder.

Jess stood up abruptly, saying, "Guess ain't nothin' goin' on around here. Let's ride."

We started to file out through the narrow kitchen. My mother ignored us as we started to leave.

"Eva, I'm gone," said Jess.

She continued to stand there stirring, her back to me. I went to her then, and put my arms around her to hug her from the back. She started to cry. I tried to look at her to say with my eyes that it was all right, everything was all right, but she turned her face and looked away in awful shame. When I released her, she hurried out of the room, crying loud and uncontrollably.

Melody saw me to the door, where she handed me a

piece of paper with her phone number on it. We didn't have to say what we both understood, that she was my half sister.

For the rest of the afternoon, Jess took me around to meet other members of the Fisher family and he took me out to get what he called the best corned beef sandwich in Cleveland. As we ate, he told me more about my mother's life and about their upbringing, the death of their mother, the violence and drinking of their father. He, too, had escaped a life of no options by joining the service. Jess and my aunts and uncles on the Elkins side were concerned that I not be mad at my mother, but I told them not to worry about that.

Another visit was arranged. I talked with Melody and got to know her a little bit. This time, my mother had advance warning. We sat for a while as I asked her questions about herself and she told me about meeting my father, about their brief courtship, about her love of music, Fats Domino, and the name Quenton. She decided to call me Q. I liked that. She was hip in a way.

Though she didn't tell me what all her life had been, I understood my mother's story, recognizing it from others I'd met in the different worlds I had passed through—poverty, loneliness, rejection. She did say that early on she tried to get this man to marry her so that she could come for me and take care of me in an environment that would satisfy social services. "But there jes' wasn't enough love there" was her reason for it not working out that way.

After giving birth to me, Eva Mae Fisher went on to bear four other children, including Melody. All five of us grew up as wards of the state. Although the other four had spent varying degrees of time with her, she was never able to raise any of them on a continuing basis. Over the years, our mother had been, for various reasons and for various periods, hospitalized, incarcerated, and on probation. According to records, she had used aliases, one of which was Eva Gardner.

Later on, we stayed in touch for a while, but as two people with only DNA in common, being so different, and given the circumstances of our status, creating a true familial relationship was not possible. I liked her, and I'm sure she liked me, too. But we were still strangers and soon my spark of curiosity extinguished. And that was that.

Before I left Cleveland, I took some time to walk through the old neighborhood. As I was heading over to Glenville, Uncle Spinoza warned me with raised eyebrows, "Be careful out there. Those ain't the same Negroes you grew up with."

He was right about some of the kids I used to know from the old neighborhood, kids who I heard had gotten into some kind of trouble, with either the law or drugs.

Some of the friends I visited appeared to be the same all right, but were stuck in a time warp. When I got to the house of one of my friends, he greeted me just as everyone had when I'd gone away to George Junior, saying, "Fish, where you been?" Like I'd gone out for a damn beer.

"Out of town," was my nonchalant reply. Or, "I was in the navy." It saved on long explanations. "How's Fat Kenny?" I asked.

"Huge," I was told. Indeed, he was huge, when I tracked him down. "Fish, where you been?" he asked, with the same lackluster energy as everybody else.

I realized finally that after all those years I'd been romanticizing my memory of them, they hadn't thought that much of me at all. The truth of the matter wasn't so much that they had changed from the way I remembered them, but that I had changed.

With all the loose ends I was tying up, you might wonder if I attempted to go by either of the Picketts' houses where I'd lived. The answer is no, I had no desire to. Later, when I got back to Los Angeles, I did track down Dwight, in jail. We talked for a long time about what he himself termed the pressure cooker. It came as a great shock that before he was incarcerated, Dwight had at one point moved down to Mississippi to live with Mr. and Mizz Pickett. He had nowhere else to go. When he went to jail, Mizz Pickett was the one sending him care packages of stamps and such. The irony to me was that Dwight, who had been so desperate to get off the island, was the one with known relatives like Uncle Mickey, where I, who stuck it out to the bitter end, had none. Now I was the one who had family and who had gotten off the island, while Dwight and Flo ended up with the Picketts.

Did I ever see or speak to any of the Picketts? Yes, in my attempt to track down pictures of myself as a child, I had a pleasant conversation with Mercy, still as good-hearted as ever, and she suggested that I speak to Teresa. All that was left of me in the Pickett family photo albums, it turned out, was a small, creased black-and-white photo of me sitting with Dwight on the front stoop of the Picketts' house. I look to myself as I felt then—an uninvited guest in the world.

In the course of my phone conversation with Teresa, she raised the issue of seeing and forgiving her mother and also her sister Lizzie, who had acknowledged feeling ashamed for treating me so badly, to which I replied by letter:

> I understand the concept, and in many ways I have forgiven those persons, for my benefit.

A shame, what could have been an extraordinary deed in her life is an ugly thing. I was a child with no responsibility for being on earth or in her home. With no way to protect myself, she had the power to crush out my spirit, and she did. She had the power to strip me of all feelings of self-worth, so she did. She had the power to beat me to dust, and she did. In my estimation, her behavior was too bizarre for her to wear the title mother, her cruelty more than criminal. . . .

I never will understand how an adult who considers him or herself decent and respectable can stand idle while a defenseless child is being hurt mentally or physically by another adult or child. No matter what the relationship to the abuser, the moral duty, above any allegiance to kinship, is to do whatever possible to stop and report the criminal and his or her indignity. If not, one certainly becomes a participant by remaining idle and silent. There is no excuse, none.

Now this calf has grown to be a bull with horns, able to protect himself from others. . . . No, I don't think it would be a good idea for me to see her, she's but a stranger I care never to encounter again. My stainless feeling of antagonism is of the seed she's sown, so comes the time to reap. . . .

Now, here I am this day, with my accomplishments, reflecting on my years with the Picketts. No credit due. Yes, the gusty winds of my melancholy youth have shifted, and brought to me a fine sunny day. I'm thankful, fortunate to have found some peace in my lifetime.

I left Cleveland as I had arrived, with Aunt Eda. We were flying together as far as Chicago, where she would disembark and I would fly on to Los Angeles. I had not cried throughout the entire time I was in Cleveland. But when the plane started down the runway, as I was concentrating on the blue lights outside my window, my protective shield came down at last and I could no longer dam up the tears. They fell like the rain that had fallen in my life, something I could never control. Eda put her hand on my back and patted it gently. I cried more and she patted me more. At O'Hare, we said good-bye, promising to see each other again very soon. Unfortunately, that didn't happen, due to a subsequent illness that would take her life two years later.

Alone on the flight from Chicago to L.A., I thought about what had happened to me, from beginning to end. I had found out the truth of who my parents were, so different from what I had imagined. They were human and flawed, like me, but I loved them and forgave them—if there was anything to forgive.

In a sense, I felt that I had been summoned to this awareness by forces unseen, by my father, Eddie, himself. This feeling was captured in a passage I later read in *Clea,* one of the four books of Lawrence Durrell's *The Alexandria Quartet:*

> Yes, but the dead are everywhere. They cannot be so simply evaded. One feels them pressing their sad blind fingers in deprivation upon the panels of our secret lives, asking to be remembered and re-enacted once more in the life of the flesh, encamping among our heartbeats, invading our embraces.

"Remembered and re-enacted." That was what it was.

For all those years of my growing up, the tidbits of my true story—the messages that came in bottles, the clues that the social workers pulled from my files—had seemed like a play, an intricately plotted script that was being enacted independently of me. The stories of Eddie and Frances and Eva and all their tragedies and the family members whose lives were intertwined with theirs, I saw them, too, as a part of the grand design of this play. It had seemed the whole time I was in Cleveland that the play was still going on, everyone stuck in their roles and having to deliver their respective lines. Only on the plane, four hours in the air, did I really realize that I was never an audience member just watching, but a player, in fact the star, the central figure of the story . . . this story of my life.

Imagine.

—John Lennon

post-memoir

FINDING
FISH

I wake to the smell of pancakes. Disoriented from sleep, for a moment I don't know where I am—still back in the dangerous caverns of memory that I've been combing lately, or here, and now, safe at home.

As the pale colors of the bedroom come into focus and I see that the golden California morning sun is beaming down its rays to put on a show just for me, I relax and settle back in the big, soft bed, marveling at the magic that is my life. Now I can hear voices from downstairs—LaNette's rich, warm laughter and Indigo's sweet, soulful singing. If there is any music more beautiful than waking to the joyful voices of my wife and daughter, I haven't heard it yet.

We eat a lot of pancakes at the Fisher house. Not every day, but not just on special occasions, either. Then again, we like to say that every day is a special occasion, a policy enforced by LaNette, whom I have named the family's official Minister of Fun. In our house, she has proclaimed, every meal is a celebration and an opportunity for giving thanks. I'm glad to go along with that. And Indigo—who has been in this world for a little over two years—doesn't know any other way.

Sometimes it seems that the life I am living now is only

a dream, that I have just conjured it all up to escape the harsh reality into which I was born and otherwise would still be living in. One could argue the point. I think back upon a childhood full of longing for belonging and see my life now as what I have created out of my dreams. An image comes to mind of Mrs. Brown at the orphanage in Cleveland, me sitting at her side, telling her, "You'll read about me someday." I was definitely dreaming then.

With no evidence of that ever being possible, I clung to that preposterous vision and with the force of those dreams willed it and made it happen. Not because I needed to be famous, but because I needed Mizz Pickett to be wrong. I needed the world that made me feel uninvited to be wrong. So I imagined myself free. I imagined myself loved. I imagined myself as somebody.

The further my boat sailed away from Shipwreck Island, the more vivid and grand and real my dreams became. Then, in 1992, when I made my trip back to Cleveland, the ultimate dream came to pass—I found my family and I found myself. I unearthed the buried treasure of who I was and had always been; the inheritance of family, of shared traits and shared blood. And I came back to Los Angeles, back to my job as a security guard at Sony Pictures Entertainment, carrying that treasure high, like a very rich man. I may have only been a security guard at the time, still shy and soft-spoken, but inside I felt as though I had returned with the Holy Grail.

And here's where the dream of my life turned rapidly and dramatically into a wild, wonderful, different kind of adventure. It seemed that after I told my supervisor my story—only because I needed him to give me the time off to go meet my family—he had been moved enough to tell other people at work about me, and then those people had been telling others my story, too. By the time I got back to work, almost every producer on the Sony lot had heard about me. Suddenly, I was being courted by several production entities all making different offers to option my life rights in order to make a movie about me.

If this part of the dream sounds too good to be true, it's because there was a catch—they all wanted an established screenwriter to write my story. Whenever I suggested myself to write it, I was politely dismissed. I may have been inexperienced in deal-making for movies, but when I was offered money for the story alone, my better instincts told me not to sell myself short. So I emerged from the flurry of excitement empty-handed.

But rather than being discouraged, I was motivated. It occurred to me that nobody had to give me permission to write. As long as I could afford paper and pencils, there was no reason I couldn't just do it. So I began writing, sometimes on the backs of my schedule on the lot, sometimes on paper towels when I ran out of ready paper.

In the spring of 1993, the dream shifted into a higher gear of reality as the result of a series of chance meetings— or so they seemed at the time. The first of these encounters was at the studio lot while I was talking with a limo driver who happened to mention a free screenwriting class at Bethel A.M.E. Church taught by Chris Smith. After attending the class one time and telling my story to Chris, he, in turn, delivered me an introduction to a producer named Todd Black. I went to meet him, a young but wise, up-and-coming producer, and told him my story. I also showed him the very rough beginnings of the screenplay I had handwritten.

Within the week, Todd called me to his office. He was on the phone when I arrived. Thinking maybe he was going to return the pages with a thanks-but-no-thanks pep talk, I stood uncomfortably until he was finished.

"Please," Todd said, after he hung up, "have a seat," and he gestured to a plush modern armchair across the desk from him.

He began slowly, saying that although he found the story very compelling and what I had written very promising, he wasn't in the position to have his company make me a deal.

"Oh," I said, barely concealing my disappointment.

"So here's what I can offer instead," Todd went on. "I know you can't write the screenplay and work a full-time job as a security guard. But if you quit your job, I could give you office space to write here. I'll pay your salary."

What he was offering took a second to sink in. He was going to pay me out of his own pocket for me to finish the script.

Echoes of lessons from the past rang in my ears. I thought of Mario Gambino telling me, "You know, Fish, opportunity comes in life only a few times. And when it does, you gotta grab it."

I grabbed the opportunity Todd gave me for all I could. Under his tutelage, I wrote and wrote and wrote. I wrote forty-one drafts. After almost a year, Todd sold the project to 20th Century Fox.

I titled my screenplay *Finding Fish*. It was not only about the discovery of myself, whom my classmates named Fish, but it was also an allusion to the biblical saying that you can give a man a fish and he'll eat for a day, but teach him to fish and he'll eat forever. I had been taught a craft. I had become the artist I'd always dreamed of being, painting with words and images to create moving pictures. Telling my own story was just an icebreaker. There was a lot more I had to say, stories upon stories that I had lived or imagined, a gold mine of riches I had unearthed inside of me, overflowing with characters and plot ideas ready for the telling. All kinds of stories, dramatic, historical, comedic, romantic, scary, and adventurous. Soon I was being hired to write those stories. The fact that I had found something that I am considered talented enough to do and earn a great living by doing it was just incredible. But the fact that it happened to be something I also love doing was even better.

As the pancake smell from downstairs becomes absolutely irresistible, I rise from the bed and begin straightening the sheets and covers before going down to breakfast. My mind travels back momentarily to some of the other visions I made up for myself—that I would become a family man, a good provider, a strong, loving husband and father, in a secure, love-filled home. And here I am, living that vision, with good neighbors and good friends.

"Daddy, pannycakes!" Indigo calls from the bottom of the steps. Unable to wait for me, she bounds up the steps, meeting me at the top and leaping into my arms, her arms wrapped around my neck as she hugs and kisses me and wishes me, "Good morning, Daddy!"

At two years and three months, Indigo is tall for her age, with my dark, serious eyes and LaNette's gorgeous mouth and smile. She is funny, smart, passionate, and determined. Slipping out of my arms, she takes my hand and hurries me along.

Together we go carefully down the steps and into the kitchen. LaNette, the prettiest woman I've ever seen, pretty as ever this morning in blue jeans and a T-shirt, is setting the table—fresh flowers, fresh orange juice, scrambled eggs, veggie sausage, and pancakes.

For a split second, I'm transported to the day of November 10, 1995, to what is known as Main Street on the lot of Sony Pictures Entertainment. After finishing up a meeting with Todd Black, I had stopped there to talk to some of my friends from security, not far from the Sony store, when out of the corner of my eye, I saw a vision of great determination walking briskly past me. She was tall, slender, with more than shoulder-length soft hair, her arms reaching out as though to speed up her stride. She reminded me of the way women walked in old MGM movies, like big blaring music ought to be sounding out and rising to a crescendo.

"Who's that?" I asked the group around me, watching as she headed off toward the Sony store. I vaguely remembered meeting her in passing in the security offices about a year and half earlier.

Carol, a security department assistant, said, "That's LaNette Canister." She went on to say that LaNette was an account manager at Culver Studios for the security company that contracted security on the Sony lot.

Before Carol could finish, I interrupted, asking, "Would you give her my phone number?" Being my typical shy self, I thought that might be better than asking to be reintroduced. I had heard people say that lightning only strikes once. What if this was lightning striking a second time? Maybe I wasn't ready before. But suddenly I felt that I was. Something about LaNette told me she was nice. Master people watcher that I was, I'd learned to spot clues that someone might be fussy or trouble or too cool or too worldly or too hurt or too something that would raise its difficult head later on.

Earlier that year I had elected to retire from the dating world, having decided that I wasn't having very much luck meeting girls who were suitable to my personality. I was tired of the whole scene. I was tired of trying to meet nice women, tired of selling myself, tired of having to convince them that I was a nice guy, tired of meeting families and trying to win their approval. But something about LaNette made me consider coming out of retirement.

What I didn't yet know was that LaNette was tired of dating, too. In fact, I later learned, she had even recently sought guidance on finding men suitable for marriage by reading a book called *Getting to "I Do."*

Not knowing any of this, I sensed an outside influence tapping me on the shoulder, drawing my attention to her. LaNette Canister seemed more than nice. She had the instant effect of making me feel the way I did when I was popular, in a simpler time when love was straightforward and true and when what I knew about girls I learned in songs by the Dramatics. I had to seize the moment.

Carol agreed to approach Ms. Canister on my behalf and went to meet her farther down the street when she emerged from the Sony store. Carol explained that a man asked that she take his phone number.

As LaNette likes to tell the story, she immediately recalled that it had been clearly stated in *Getting to "I Do"* that a woman with hopes of getting married must never call a man, so she responded, "I don't call men."

Then, this is the part LaNette really likes to tell, she looked in my direction and felt her eyes bulging, as if to say, "*Boy-yo-yo-yoing!*"—like in a romantic cartoon after Cupid's arrow has been shot. I like that. I like it when she says that the phrase *ebony god* came into her mind. She decided that even though she didn't call men, she could certainly approach me and introduce herself. After she did, LaNette asked if I'd mind walking her to her car. She was double-parked, she said, because there were no spaces available and she had to price a cordless phone at the Sony store for her boss, which explained her earlier hurry.

On the way to her car I expressed my concern that if LaNette refused to call men, how were we going to talk? Her solution was to give me her phone number.

The same evening I called LaNette and we talked at length. Born and raised in Compton, she was the youngest of three children of devoted parents, Sandy and Gwen Canister. Growing up with love, encouragement, and belief in her potential, she had the confidence to apply and be offered a full scholarship to Milton Academy, a boarding school in Milton, Massachusetts, then ranked the third-best private high school in the nation. After graduating from there, LaNette went on to get her college degree at the University of Pennsylvania.

"Pennsylvania?" I joked when I heard that. "I went to school in Pennsylvania!" Then I proceeded to tell her about George Junior, not exactly Ivy League, but a place where I had learned a great deal nonetheless.

When I discovered what an avid reader she was, I asked if she would be interested in reading any of my writing.

"I'd love to," LaNette said, without hesitation.

Scripts in hand, I arranged to meet her on a park bench under the big clock in front of the Capra Building. In that old-fashioned, romantic setting that looked to be Anywhere, USA, I sat down next to LaNette, thinking that she was prettier than I had even noticed before. Dressed neatly in a tailored navy blue business suit, with stockings and matching shoes, she was so soft and lovely. What most captivated me were her amazing light brown eyes.

When I walked her to her car, I continued to notice them. They told me everything I needed to know about her. She was a decent girl, I decided, a good person. She was refreshing—shy in her own way, not shy like me, but in that context she was, touchingly so.

362

"You have beautiful eyes," I felt emboldened to say.

"Thank you." LaNette smiled.

It wasn't the color of brown that made them so special, nor the clarity of the whites, nor the way the light reflected so brightly in them. What was so beautiful to me was that in those eyes I found myself looking at someone who was missing her other part. It wasn't a sad or a bad thing; we all come into this world without our other part and hopefully we find it at some time. I thought because of what I saw in her eyes, that maybe it was her time. Not that I felt that I was the one, but I felt it was missing for her. What I felt for me, before I saw myself in her life, was that this day had brought forth a seed—a seed that could grow into something.

Eased into the soil of a friendship that flourished in daily phone calls that lasted hours on end, by Thanksgiving of 1995, the seed had taken root in true love. After that, not one day went by that we didn't see each other, until I think we both came to the conclusion that we had always been in each other's lives, only waiting until we were ready, as we were now, to be together.

A year later, a week after Thanksgiving, LaNette became my wife. To this day, she claims it happened because she used all the techniques she learned in *Getting to "I Do."* As for myself, I give credit to Nat King Cole— one of the greatest, most romantic artists of all time, an artist I had only discovered just before my courtship with LaNette began and whose songs were the theme for our getting together.

In the dream that kept getting better and better, we soon learned that LaNette was pregnant, due sometime around Thanksgiving of 1997.

After a walk in a nearby park one evening toward the end of her third trimester, I went into the bedroom to find

LaNette on the rocking chair. She was wet and the floor underneath her was wet.

"I think my water broke," she said nervously.

Without saying a word, I instantly went to work cleaning up. Out came the vacuum cleaner, the laundry basket, and the detergent.

"Antwone, what are you doing?"

"Cleaning," I explained, not sure exactly why that was my response to the baby's imminent arrival, except that it was what I would have done when any important guest was due to arrive. I didn't want the baby to come home and have it not be perfect. Everything was so out of order when I was born, I wanted the opposite for my child. I wanted everything in order, a welcoming environment, a clean, loving home. Not that a dirty sock in the hamper should have mattered, but that, too, I had to wash before heading off to the hospital with LaNette.

Labor lasted eighteen hours. LaNette was magnificent. I'll never forget my first sight of my daughter as the doctor held her up and she began to cry. She was amazing. This was a miracle. I felt like saluting everybody in the room—the doctors, the nurses, my mother-in-law, and, of course, LaNette. In all my travels, I had never seen anything like this.

When the doctor put the baby on LaNette's chest, I went and leaned down over LaNette's shoulder. Incredible. Moments before this little person hadn't existed and suddenly there was a new person in the world, the person we had been waiting for.

The baby was crying, breathing in her first breaths, and LaNette said something to her and she stopped crying and raised her head to look at the bearer of the familiar voice. She knows her mother, I thought in total wonderment. She knows her.

Then it was my turn. After I had the honor of cutting the umbilical cord, I followed as a nurse took the baby over to another room to bathe and weigh her.

LaNette called to me, "Antwone, you stay with that baby!"

"Don't worry, I will," I called back, thinking I wouldn't let our baby out of my sight ever, "I got it!"

At the sound of my voice, the baby stopped crying and she lifted her head up at me. I'd been singing the song "Summertime" to her during the whole pregnancy and to welcome her into the world with that familiar number, I started to sing. *"It's summertime and the living is easy. Fish are jumpin' . . ."* We were Fish, after all. *"And the cotton is high."* I could swear that my newborn infant daughter smiled, knowing that I was her father, just before starting to cry again as I sang, *"Your daddy is rich and your mama's good looking, so hush little baby, don't you cry."*

LaNette and I chose the name Indigo because we liked the sound of the name and that it had been used for a Duke Ellington composition, "Mood Indigo," and because we looked at our beautiful, sweet little baby girl and knew that's who she was.

With pancake crumbs on her smiling mouth, Indigo reaches across the table to pat my hand, like she knows what I'm thinking about.

LaNette and I clear the table together as Indigo requests to watch her latest favorite video, *Annie*. In the family room, I put it in the VCR and she snuggles up on the sofa, beckoning me to join her. I have work to do in my office, but how can I say no? Even if it does mean watching a story about orphans being mistreated.

Promising that I can watch only until Indigo's favorite song, "It's a Hard-Knock Life," comes on, I reflect how fatherhood has given me back the special powers of imagination that I had lost so long ago. The ability to watch the sun's light show on the wall has come back to me and I've even started to believe that I can melt snow, as I did so very long ago. Fatherhood has allowed me to be young again, to feel innocent once more and not so used up.

Maybe I wasn't lucky enough to be invited into the world, but LaNette and Indigo make me feel welcomed nonetheless. A space has opened up inside, a space that was perhaps reserved for having a child. I am a brand-new me—with a chance to live it over, to give Indigo the childhood I didn't have, to be a great father, to be her friend, to keep her safe and secure, to warn her of danger, to teach her and love her.

On cue, Indigo sings out, *"It's the hard-knock life . . . for us!"* and jumps up from the sofa to dance along with the girls in the movie.

LaNette comes in and she and I laugh at Indigo's exuberance.

The irony is that for Indigo, a world where people get kicks instead of kisses and tricks instead of treats is only make-believe. One day, when she's older, she'll know the story of my childhood. She'll understand why it is I am such a protective parent, why after spending so long finding a family, I find it difficult to be separated from her or her mother for even a night, and why, whenever we are separated, it's never "good-bye" but "see ya later."

I hope that in knowing my story Indigo will also understand what I have learned: that nothing happens to us by chance. My course was not determined by the whims of cruel weather, it was set carefully and precisely. Everything that happened did so for a reason, at exactly the right time, in exactly the right way.

I wish I could say that the past doesn't hurt me still. But the truth is that delving into the past can be very painful, that I have scars, and that I will always have scars. On the other hand, when I think of Dwight and Flo and Keith, I know that I am among the most fortunate. When Indigo knows my story, I hope she sees that and understands I was made of the same strong stuff of which she is made. I hope she sees my fortune as the result of the true goodness of people who exist in this world.

I'm thinking of people in my life to whom I will always be grateful—people like Mrs. Profit, my teacher who believed in my potential, even when there was little outward indication of it.

I'm grateful to all the teachers and guides who were there in my life at the necessary time—to my social workers, like Ms. Edwards, Ms. Nees, and Bill Ward; to the great state of Ohio and its taxpayers; to everyone at George Junior Republic; to Chief Lott, Chief Akiona, Commander Williams, Captain Anderson, and the United States Navy; and to my good friend Todd Black. I am most grateful to a woman named Mrs. Strange, my first foster mother, whom I have never had the chance to meet and thank, in whose home, in the first two years of my life, I learned to love pancakes.

acknowledgments

To Mim Eichler Rivas, thank you for your guidance and structure through this journey. You're a beautiful spirit. Jennifer Rudolph Walsh, every time I hear your voice I feel as if I am under the umbrella of protection. You are wonderful. We'll do another one. Henry Ferris, thank you for helping to make a dream come true. Thank you to everyone at William Morrow for your thorough publishing care and good taste. To Virginia Barber and everyone at The Writer's Shop, thank you very much. Rob Weisbach, thank you for hearing me and responding. Jeff Frankel, what would I do without you? Probably nothing. Thanks, man. Thanks to Colden McKuin & Frankel. I feel connected. Endeavor, you are the convalescent home for frustrated writers. Thank you. I'm better now. Matt Lichtenberg, it is easy to get people to walk with you in the cool of the evening, but not many will walk with you in the heat of the day. You are a good friend. Thanks to everyone at GLWG for taking care of us. Chris Smith, thanks for pointing me in the right direction to a great future. To Todd Black, thanks for teaching me to fish. Jason Blumenthal, thank you, my brother. Thank you for everything. Randa Haines, you are magical. Ruth Black, thank you for being an ear during my hours of frustration. Denzel Washington, what can I say? I trust you with my life. Thank you. Joe Pichirallo, I feel safe that you are around because you actually "get it." Robert Simonds, thank you for all the opportunity and trust. Pauletta Washington, thanks for believing. It always helps. Rita Pearson, you understood me from the start. Thank you.